AIRBORNE GEOPHYSICS AND PRECISE POSITIONING

SCIENTIFIC ISSUES AND FUTURE DIRECTIONS

Committee on Geodesy
Board on Earth Sciences and Resources
Commission on Geosciences, Environment, and Resources
National Research Council

NATIONAL ACADEMY PRESS
Washington, D.C.
1995

Support for this study was provided by the National Science Foundation, Air Force Office of Scientific Research, National Aeronautics and Space Administration, Defense Mapping Agency, and the U.S. Geological Survey.

Library of Congress Catalog Card No. 94-68678
International Standard Book Number 0-309-05183-5

Copies of this report are available from:

National Academy Press
2101 Constitution Avenue, N.W.
Washington, D.C. 20418

800-624-6242
202-334-3313 (in the Washington Metropolitan Area)

Cover art by Shelley Myers, Project Assistant, Committee on Geodesy, Board on Earth Sciences and Resources, National Research Council. Myers's work is exhibited widely in the Washington, D.C. area and has won several area awards. The cover depicts an airborne geophysical survey.

Printed in the United States of America

COMMITTEE ON GEODESY

The National Academy of Sciences is a private, nonprofit, self-perpetuating society of distinguished scholars engaged in scientific and engineering research, dedicated to the furtherance of science and technology and to their use for the general welfare. Upon the authority of the charter granted to it by the Congress in 1863, the Academy has a mandate that requires it to advise the federal government on scientific and technical matters. Dr. Bruce Alberts is president of the National Academy of Sciences.

The National Academy of Engineering was established in 1964, under the charter of the National Academy of Sciences, as a parallel organization of outstanding engineers. It is autonomous in its administration and in the selection of its members, sharing with the National Academy of Sciences the responsibility for advising the federal government. The National Academy of Engineering also sponsors engineering programs aimed at meeting national needs, encourages education and research, and recognizes the superior achievements of engineers. Dr. Robert M. White is president of the National Academy of Engineering.

The Institute of Medicine was established in 1970 by the National Academy of Sciences to secure the services of eminent members of appropriate professions in the examination of policy matters pertaining to the health of the public. The Institute acts under the responsibility given to the National Academy of Sciences by its congressional charter to be an adviser to the federal government and, upon its own initiative, to identify issues of medical care, research, and education. Dr. Kenneth I. Shine is president of the Institute of Medicine.

The National Research Council was organized by the National Academy of Sciences in 1916 to associate the broad community of science and technology with the Academy's purposes of furthering knowledge and advising the federal government. Functioning in accordance with general policies determined by the Academy, the Council has become the principal operating agency of both the National Academy of Sciences and the National Academy of Engineering in providing services to the government, the public, and the scientific and engineering communities. The Council is administered jointly by both Academies and the Institute of Medicine. Dr. Bruce Alberts and Dr. Robert M. White are chairman and vice-chairman, respectively, of the National Research Council.

PREFACE

The Committee on Geodesy (COG) of the National Research Council (NRC) has long been involved in evaluating the opportunities for new insights and applications that accurate geodetic measurements offer the earth, ocean, and space science communities. With the development of the Global Positioning System (GPS) and the possibility of having very precise aircraft navigation, new scientific opportunities exist for high-resolution surveys of the topography and gravity field of the Earth. To assess these new opportunities and the ways in which airborne techniques using GPS might complement other ground- and space-based techniques, COG held a workshop on airborne geophysics in July 1993. The goals of the workshop were as follows:

- to evaluate the current status of airborne geophysics and precise positioning techniques;
- to identify the scientific objectives that are made possible by integrating aerial measurements and precise positioning techniques; and
- to propose strategies for expanding and optimizing the future development of airborne research platforms, aircraft-based measurements, and precise positioning.

A steering committee of six individuals from COG with expertise in both the instrumentation and the scientific aspects of airborne geophysics planned the workshop. Steering committee members were Robin Bell (chair), Thomas Herring, Christopher Jekeli, J. Bernard Minster, Richard Sailor, and David Sandwell. Held on July 12-14, 1993, in Washington,

D.C., the workshop consisted of plenary and poster sessions that allowed interaction between federal managers and the applied and basic research communities.[1] Workshop participants (see Appendix D) included approximately 40 scientists and engineers who are active in the areas of airborne geophysics and precise positioning, or who are addressing research questions that could benefit from the application of a new generation of airborne geophysical measurements.

The workshop (see the agenda in Appendix C) was organized around four principal technical elements of airborne geophysical measurements and the scientific objectives that can be addressed by these new techniques. The themes of the workshop were as follows:

- GPS positioning;
- integrating GPS and Inertial Navigation Systems (INS);
- recovery of topography from aircraft; and
- airborne gravity field measurements.

The scientific objectives of the next generation of airborne geophysics programs were discussed within the framework of the technical elements. Working groups chaired by members of the steering committee developed conclusions related to these themes. The major conclusions of the working groups are presented in Appendix B.

The following report is based primarily on information presented at the workshop and on the expertise of the workshop participants and the steering committee. The Committee on Geodesy worked with the steering committee in the preparation of this report and takes full responsibility for its conclusions and recommendations.

Robin E. Bell, *Chair*
Steering Committee on Airborne
Geophysics and Precise Positioning

---

[1] Abstracts of the papers presented at the workshop are available from the Committee on Geodesy, National Research Council.

CONTENTS

AIRBORNE GEOPHYSICS AND PRECISE POSITIONING

SCIENTIFIC ISSUES AND FUTURE DIRECTIONS

EXECUTIVE SUMMARY

INTRODUCTION

Since the *Challenger* circumnavigated the globe on its oceanographic survey in the mid-1870s, scientists have mounted extensive interdisciplinary expeditions to study large regions of the Earth. The ability to probe the Earth from remote platforms has resulted in numerous scientific advances, such as the revolutionary images of previously uncharted ocean floor that were produced by marine geophysical technology in the 1960s. These images eventually helped to trigger a major paradigm shift in the earth sciences with the development of a fundamental theory of modern geophysics, plate tectonics.

Scientists' understanding of geophysical processes (including lithospheric, oceanographic, and ice sheet processes) has been limited by their inability to make accurate, precisely-positioned measurements. Many of the measurement tools used historically by earth scientists for regional studies are not sufficiently accurate to model physical processes and thereby to improve the understanding of natural hazards and the distribution of nonrenewable resources. Physical barriers, such as inaccessibility by land or the impediment of a hazardous environment, and limited resources that prevent the surveying of large areas by conventional means, pose other difficulties. Maps of geophysical measurements commonly delineate political and physiographic boundaries rather than geological trends. For example, gravity coverage in Africa is sparse in those regions that are politically unstable; this distribution makes the geophysical interpretation of the gravity data within a global framework difficult, if not impossible.

Collecting data from satellites or aircraft can overcome these problems in sampling strategies. Satellites provide global coverage, but they may

1

not provide data with sufficiently high resolution (i.e., the ability to discriminate features spatially), or they may require long lead times between design, launch, and data acquisition. Airborne platforms provide an attractive alternative for studying the Earth. Airborne magnetic measurements have long been available to the earth science community and are important for mapping regional geology, identifying regional deformation patterns, studying seismically active faults, and finding mineral and petroleum resources (e.g., NRC, 1993). Despite the appeal of airborne measurements, their widespread use for other applications, such as gravimetry and precise topographic mapping, has been hampered by the inability to position aircraft accurately. Measurements of gravity and terrain have little value if the aircraft obtaining these measurements cannot be positioned in three dimensions to better than 10 meters (m) horizontally and 1 m vertically. High-resolution surveys that require closely-spaced flight lines have also been difficult to carry out successfully because of problems with positioning the aircraft.

The development of the Global Positioning System (GPS) in the past decade enhanced the ability both to navigate and to position an aircraft. GPS consists of 24 satellites that can be used to position an aircraft to decimeter accuracy at any point on the globe using differential techniques. Differential techniques exploit the concept that errors in the GPS signal propagation and timing can be corrected by using a receiver in a fixed location as a reference for the moving receiver. Prior to the development of GPS and its commercialization, precise positioning was possible only with a great investment of money and manpower to install a local land-based navigation network. The airborne measurement platform combined with a precise positioning capability such as GPS holds great potential for new insights and new applications in solid earth geophysics. Although great strides are being made in this direction, the full extent of the potential for widespread application has been largely unexplored.

The purpose of this report is to highlight the advances, both potential and realized, that airborne geophysics and precise positioning have made or can make possible to the solid earth sciences. The report first discusses the state of the art in airborne geophysics as integrated with new precise positioning systems, focusing first on the new technology to map topography and the gravity field and then on recent advances in precise positioning. These technologies are described in Chapter 1. Chapter 2 then outlines the scientific goals of a focused effort in airborne geophysics, including advances in our understanding of solid earth science, global climate change, the environment, and resources. Chapter 3 identifies the technological advances in measurement, positioning, and aircraft design that will be required to aggressively pursue the scientific goals discussed in Chapter 2. The recommendations of the Committee on Geodesy for achieving the goals and directions outlined in the report are presented in Chapter 4.

SUMMARY OF RECOMMENDATIONS

At present the geoscience community has insufficient access to airborne technology to fully realize its potential for studying the Earth. Although airborne platforms exist in many agencies and organizations, their use is commonly restricted to employees and affiliated scientists. In other cases, the complexity, expense, and logistical difficulties of conducting an airborne geophysical campaign may deter scientists from fully utilizing this important geophysical tool. There needs to be a dedicated airborne earth science facility, consisting of aircraft, instruments, and personnel, that any scientist can propose to use. Such a facility could be managed by any of a variety of university, government, non-profit, and/or private-sector coordinating bodies.

Recommendation 1. The new capabilities in precise positioning and accurate navigation should be made more accessible to the user community. An initiative to increase access should include both the establishment of an airborne earth science facility and a coordinated effort at educating its potential users.

Most airborne missions are equipped with only the instruments that are necessary to meet the specialized goals of the operating organization. If these organizations were to coordinate some of their missions, a great deal of additional data could be collected at a relatively small incremental cost. The collection of multiple data sets, each of which is collected for a specific research objective, would foster interdisciplinary research and make better use of the aircraft.

Recommendation 2. Airborne geophysical measurements should be coordinated across disciplines, programs, and funding agencies to promote interdisciplinary research and to optimize use of the aircraft.

Accurate measurements are needed for both long-wavelength regional studies and short-wavelength process-oriented studies. Some of these studies require multiple measurement systems on a single platform. Future technological developments must address all of these needs to ensure that airborne geophysical methods will be practical for the scientific, resource, and environmental industries.

Recommendation 3. To ensure uniform coverage that is sufficiently accurate to resolve both long- and short-wavelength geologic features, technological developments should aim at integrating GPS with a broad spectrum of well-calibrated measurement systems.

The geophysical applications discussed in the report require access to GPS signals that make it possible to locate the aircraft's antenna to a few cm. These applications, as well as future advances in airborne technology, are being hampered by the Department of Defense's implementation of Antispoofing, which encrypts the P-code for reasons of national security. It is possible for the geophysical community to compensate for the effects of Antispoofing, but the necessary methods are expensive or are still in the developmental stage.

Recommendation 4. In light of the serious impact on airborne geophysics, particularly for emerging industrial applications, the continuous operation of the Antispoofing system should be carefully evaluated.

1

AIRBORNE GEOPHYSICS: A POWERFUL TOOL FOR STUDYING THE EARTH

INTRODUCTION

Since the early days of balloon photography and military reconnaissance, people have been struck by the broad view of the Earth that the airborne perspective provides. From the first photographs of Earth taken from aircraft with hand-held cameras to the highly refined swath terrain profiling systems now being developed, aircraft have provided a unique approach to studying earth science. Compared with ground- or space-based methods, airborne techniques offer the advantages of improved access, rapid sampling at scales that are optimal for many geophysical problems, and a tremendous potential for interdisciplinary studies at intermediate scales.

Access

Aircraft provide the capability of traversing regions that are otherwise difficult or impossible to cover. Examples include remote areas of the Rocky Mountains, the treacherous waters of the Drake Passage in South America, and the thickly-vegetated Amazon basin. The map of the Trinidad Quadrangle, Colorado, demonstrates the advantages of an airborne program: the land-based gravity survey is restricted to the passable roads and stream valleys in the region (Figure 1.1), whereas an airborne survey could systematically sample the entire region.

5

6

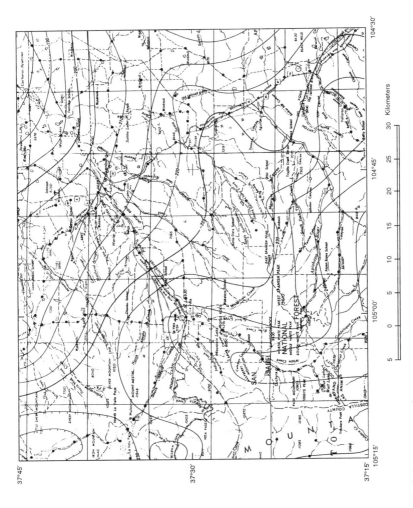

FIGURE 1.1 Gravity map of a portion of the Trinidad Quadrangle, Colorado. Note that the locations of gravity measurements are clustered near the roads and in small stream valleys. Contour interval is 5 mGal. (Figure modified from Peterson et al., 1968. Courtesy of P. Hill, U.S. Geological Survey).

Areas that are physically accessible but that have social, economic, or political barriers are also potential candidates for an airborne survey. Where land access is restricted in environmentally hazardous areas because of health risks or industrial or military interests, the geochemical signature of these regions can be mapped remotely. A recent example is an airborne survey over a strip mining region in Czechoslovakia that mapped the extent of uranium contamination in the local river system (Figure 1.2).

Sampling

Proper sample spacing is crucial for obtaining meaningful scientific results. Undersampling of a region may not allow resolution of the major features that control the system. Oversampling is inefficient and may restrict the extent of the area to be studied, but may be necessary to isolate the signal from the noise. The sampling strategy chosen depends on the objectives. For example, magnetic anomalies that constrain rates of seafloor spreading can be adequately resolved with 20-kilometer (km) line spacing, aligned orthogonal to the ridge crest. In regions where plate motion is complex, a much closer line spacing (on the order of 4 km) with a complementary set of crossing lines is required to identify small-scale features and rotated crustal blocks. With airborne techniques, the scientific objectives of a study, rather than restrictions of access, allow the optimal sampling strategy to be selected.

The improved sampling that is possible with an airborne program has particular advantages in the study of the potential fields of magnetics and gravity. Many of the errors and uncertainties that arise in the interpretation of land-based studies result from removing the effects of noise. For example, humanly-generated magnetic fields introduce "social" noise in magnetic studies, and surrounding, but poorly known, topography can introduce noise in gravity studies. Both of these measurement types are more easily interpreted if they are taken a systematic distance from the source. To achieve optimal results, airborne gravity is typically collected at a single altitude, whereas airborne magnetic surveys are generally flown at a constant elevation above the terrain.

FIGURE 1.2 Environmental Monitoring System airborne gamma ray radiometrics survey of Czechoslovakia's Mimon Uranium Mine (northeastern area) showing uranium anomalies along the local river system. The brighter areas indicate high ($\geq$9 parts per million) uranium concentrations. The survey was flown at a height of 80 m and a line spacing of 250 m. The data are overlain on a SPOT satellite scene of the area so that ground features can be identified easily. (Figure courtesy of B. Larson, World Geoscience Corporation).

Airborne platforms also offer the advantage of rapid sampling rates. A recent gravity survey of Gabon in western Africa illustrates the speed in which an airborne survey can be conducted. The area surveyed was a combination of mangrove swamp, shallow water, and jungle. Using airborne gravity measurements, an area of 150,000 square kilometers (km^2), about the size of Florida, with lines spaced 6 km apart, was surveyed in 50 flights, conducted over 55 days (see Figures 1.3(a) and 1.3(b)). The final product, a detailed gravity map and geologic interpretation, was delivered to the contracting petroleum company less than one month later. A similar land-based survey would have required many more months of effort, as well as a large team of surface parties with diverse instruments and equipment to conduct the survey in areas with a variety of vegetational cover.

Interdisciplinary Studies

An important advantage of airborne survey systems is their potential for a flexible interdisciplinary platform. In the 1960s and 1970s, research groups collaborated in a systematic survey of the ocean floor using deep-water research vessels. The strategy was not simply to collect topographic measurements, to sample the ocean floor with cores, or to probe the structure of the ocean water column. Rather, these vessels provided a center for integrated earth science research. This approach has been the standard for oceanographic vessels since the *Challenger* expedition in the mid-1870s, which circumnavigated the globe and gathered a broad range of samples and observations. Alone, any one of the sample collections from this expedition would have been interesting, but together they provided a powerful stimulus for rethinking the structure and habitats of the globe. Aircraft can provide a similar centerpiece for interdisciplinary research.

The capabilities aircraft offer for such research can be exemplified as follows. In many regions, the seepage of saline groundwater into agricultural areas adversely affects crop production. In order to determine groundwater flow patterns, aircraft could carry instrumentation for both airborne magnetic capabilities and electromagnetic mapping. The airborne magnetic data would define the geologic structure, and the electromagnetic data would delineate the extent of saline groundwater. Together these data

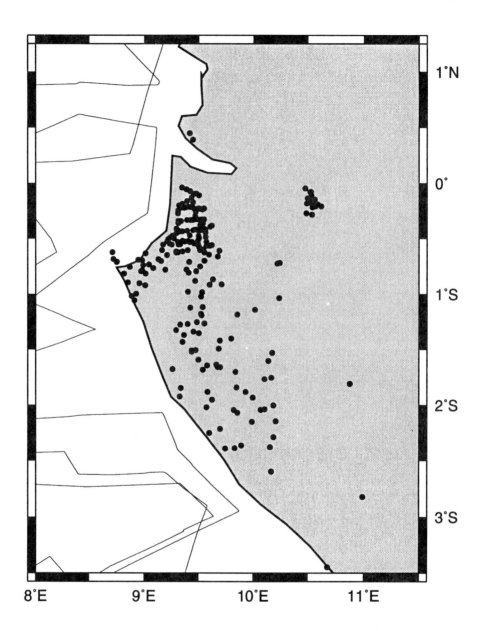

FIGURE 1.3(a) Land gravity measurements (dots) and marine gravity surveys (lines) of coastal Gabon. The land measurements were made principally on roads close to major cities. (Data from Watts et al., 1985).

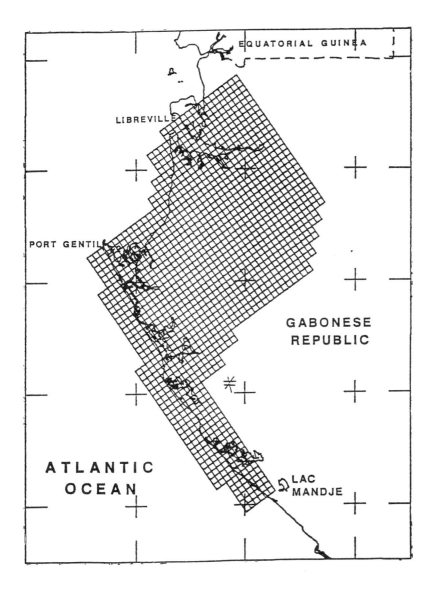

FIGURE 1.3(b) Airborne gravity survey (lines) flown over Gabon in a 6 × 6 km grid. The sampling strategy is not limited by access to roads. (Figure courtesy of B. Gumert, Carson Geophysical).

may indicate, for example, that the flow patterns are controlled by the presence of a series of igneous dikes that cut across the area. It is this capability to carry and utilize multiple instruments on an aircraft that provides new insights into the linked processes that control the earth's systems. The major drawback of using multi-instrumented aircraft, however, is the need to compromise individual measurement types. As noted above, the optimal altitude for collecting airborne gravity and magnetic measurements may differ. Nevertheless, in many instances such interdisciplinary efforts, coupled with new high-resolution techniques, offer increasing opportunities for airborne geophysics. Precise positioning, based on GPS technology, will be increasingly necessary to advance these new applications.

PRECISE POSITION CAPABILITIES
AND REQUIREMENTS

Airborne geophysical experiments require that the location of the measurements be accurately known and, in some cases, that the orientation of the platform from which the measurements are taken be well constrained. Accurately locating the experiment is the real-time, or navigational, problem. In large-scale regional studies, real-time navigation is of minimal concern, but as the applications of airborne geophysical techniques become increasingly detailed and as the instrumentation becomes increasingly sensitive, navigation issues become more significant. Navigation issues are critical, for example, in monitoring changes in topography after an earthquake. The scientific objective of understanding the deformation of the Earth requires that the same track be flown both before and after the seismic event.

Recent applications of pseudosatellites have produced encouraging results in very precise real-time applications and could have application in post-processing. Most geophysical applications require post-processing of the data because of our inability to hold the aircraft to within a meter. In general, postmission positioning requirements are more stringent than are real-time navigation requirements, and thus, more sophisticated algorithms are used for aerotriangulation applications, for example, or in computing aircraft acceleration for gravity applications.

Recovering the aircraft orientation is important to many applications, as this information may be required to correct for aircraft motion during data acquisition. Examples are the correction of laser ranges to the vertical and also the rectification of imagery.

The accuracy required for navigation, positioning, and attitude varies among operations, but, in general, the exploitation of data obtained from airborne sensors increases with improved positioning and navigation capabilities. GPS is a powerful tool for determining aircraft positioning and navigation and is potentially useful for recovering orientation. The system has the advantages of being globally accessible, little affected by local weather conditions, and having relatively low user costs. Table 1.1 summarizes the achievable accuracies for position and orientation using GPS.

The GPS system is based on a constellation of 24 satellites that transmit biphase encoded signals at two frequencies, denoted by L1 (1.575 gigahertz [GHz]) and L2 (1.227 GHz). Two basic codes are written on the GPS carrier signals at two frequencies: the coarse/acquisition code (C/A code) and the precise code (P-code). The system was designed to provide navigation to an accuracy of approximately 10 m with the P-code, and approximately 100 m with the C/A code (U.S. Department of Transportation/U.S. Department of Defense, 1992).

Researchers accessing the full range of signals transmitted by the GPS satellites have expanded the capabilities for positioning moving vehicles (such as aircraft) to the decimeter level using differential techniques. The three main types of observables that are used in the analysis of GPS data to position objects precisely are (1) the pseudorange, (2) the carrier phase, and (3) the Doppler shift. Pseudorange measurements are made by determining the difference between the arrival time of a GPS signal (as measured on the receiver clock) and its transmission time (as determined by the satellite clock). By simultaneously observing four or more satellites, it is possible to determine the position of the receiver and to correct for differences in time between the receiver's clock and the GPS satellite time system. When Selective Availability (SA) is turned on, there is high-frequency "dithering" of the satellite clock by up to 0.2 microseconds (μs). This dithering cannot be recovered from the ephemeris message without access to a classified decryption key (see APPENDIX A). This noise introduces 60-m errors into the pseudorange solutions with periods of several minutes.

TABLE 1.1 Summary of GPS Positioning and Attitude Accuracies

Model	Distance from Reference Receiver	Achievable Accuracy
Pseudorange point positioning*		100 m horizontal 156 m vertical
Carrier-smoothed pseudorange differential positioning	10 km	0.5 - 3 m horizontal 0.8 - 4 m vertical
	500 km	3 - 7 m horizontal 4 - 8 m vertical
Carrier phase differential positioning	10 km	0.03 - 0.2 m horizontal 0.05 - 0.3 vertical
	50 km	0.15 - 0.3 m horizontal 0.2 - 0.4 m vertical
Attitude determination	1 m separation 5 m separation 10 m separation	18 - 30 arcminutes 4 - 6 arcminutes 2 - 3 arcminutes

* Selective Availability on, position dilution of precision (PDOP) ≤ 3, 2 times distance-root-mean-square (DRMS) to 95 percent probability (U.S. Department of Transportation/U.S. Department of Defense, 1992).

Carrier phase measurements are made by reconstructing the carrier signal, the fundamental frequencies of L1 (1.575 GHz) and L2 (1.227 GHz). This process includes removing the biphase encoding and measuring the phase difference between the reconstructed carrier phase and a local oscillator within the receiver for both the L1 and L2 frequencies. As this carrier-beat phase rotates through cycles, the number of cycles is accumulated. Thus, the phase measurement is the accumulated phase change from the time the satellite is acquired by a receiver (locked on) until the receiver loses the signal (loss of lock). The carrier phase noise is a few millimeters; however, it also contains an ambiguity that must be determined to exploit this accuracy.

The Doppler observable is most often determined from the time derivative of the carrier phase. The Doppler observable is rarely used for geodetic positioning, but it can be useful for connecting phase measurements across small gaps (< 5 s) when the signal from a satellite is lost (e.g., during banked turns or rapid accelerations).

Together these three observables provide a powerful suite of tools for determining the time and position of a vehicle anywhere around the globe. The principal errors in a GPS position arise from errors in timing and errors introduced as a result of the propagation of the signal from the satellite to the receiver. The SA signal is a clear example of an artificially induced error that affects the pseudorange measurements and thus position determinations. The SA error size is much larger than the timing errors that are intrinsic to the GPS clocks. The propagation errors are associated with dispersive delay in the ionosphere, atmospheric delays, and the signal reflecting off objects surrounding the antennas before it is recorded by the receiver (multipath). Differential techniques based on multiple satellites and receivers are used to minimize many of these errors. Further information on GPS measurements, errors, and processing techniques is given by Wells et al. (1986), Hofmann-Wellenhof et al. (1992), Seeber (1993), Cohen et al. (1994), and Enge et al. (1994).

Differential techniques are based on the concept that many of the errors in timing and propagation can be eliminated by using a reference receiver in a fixed location and positioning the moving receiver relative to this reference. This configuration assumes that systematic errors seen by both receivers can be removed by differencing the signal. The accuracy of differential positioning decreases with increasing baseline lengths. This approach can be applied to both pseudorange and carrier phase measurements. Differential pseudorange positions are accurate to 0.5 to 3.0 m, depending on the receiver type used, and are being used experimentally for commercial aircraft systems. Differential carrier phase measurements are as accurate as 2 to 20 centimeters (cm) and generally require extensive postmission analysis, although real-time systems are under development (e.g., Frodge et al., 1994).

Several efforts have been made to demonstrate the robustness of the differential GPS approach for accurately locating aircraft. As the most difficult component to constrain is the vertical position, the recovery of this component is generally used as a benchmark. A ranging system that provides an independent constraint on the aircraft height is used in combination with the precise GPS positions. In experiments over water or ice, the position of the aircraft recovered from the independent, and probably more accurate, ranging system is within 20 to 50 cm of the elevation measurements recovered from the differential GPS carrier phase measurements (Blankenship et al., 1992). Other airborne tests have been conducted using inverse photogrammetry in which the camera positions are independently determined from aerotriangulation with ground control and then are compared to the estimated GPS positions (e.g., Cannon, 1991).

MEASURING THE EARTH'S GRAVITY FIELD

A striking example of the benefits of precise positioning of a research aircraft is the determination of the Earth's gravity field. Accurate knowledge of the Earth's gravity field is of major importance in geodesy and geophysics. In geodesy, the gravity field defines a reference, the geoid (or mean sea level under ideal conditions) for topographic heights (Heiskanen and Moritz, 1967). The geophysical utility of surface gravity data lies in constraining the density structure of the Earth. Furthermore, the Earth's gravity field is one of the few quantities (magnetics and topography being the others) that contain the same levels of spatial detail as the Earth's density, and can be measured and mapped at all these levels by a variety of techniques. Gravity measurements have applications over a broad spectrum of scales, ranging from tens of thousands of kilometers to less than 1 km (Table 1.2). The accuracy needed, corresponding to the strength of the signal, ranges from parts per hundred thousand to better than a part per million (e.g., NRC, 1990). With just a few global earth parameters, the gravity field can be modeled to an accuracy of a few parts per 10,000; the remaining structure on global, regional, and local scales must be measured.

On land, measurements of relative gravity historically have come from static deployments of the pendulum, the torsion balance, and (since 1950s, exclusively) the spring gravimeter (Torge, 1989). Coverage with these types of instruments is limited inherently by geographic accessibility, a stable environment, and human and economic resources. However, given the resources and sufficient time, land gravimetry, and, to some extent, sea gravimetry, could be measured at high resolution over many parts of the

TABLE 1.2 Gravity Accuracy and Resolution Needed for Solid Earth Science Applications (Compiled from NRC, 1990; input from 1993 Workshop on Airborne Geophysics).

Feature	Accuracy (mGal)	Horizontal Resolution (km)
Plate Boundaries and Intraplate Deformation		
Large-scale flexure	5	50
Rifts	3	10
Diffuse extension	2	5
Mountains	3	5
Sub-ice topography	2	10
Volcanology		
Volcano morphology	1	10
Volcano dynamics	0.5	5
Mineral Exploration		
Sedimentary basins	1	3
Salt domes	0.5	1
Mineral prospects	0.1-2.0	1-10

world. In fact, large gravimetry data bases exist for most of the world's land areas, although coverage is not uniform. Some parts of Africa, South America, Asia, and the polar regions are characterized by no surface data at all. The global structure of the gravity field can be discerned by tracking an Earth-orbiting satellite from several ground stations and observing the orbital perturbations caused by the Earth's asymmetric mass distribution. Today, with the accumulation of tracking data from numerous satellites and with constantly improving tracking techniques, the Earth's gravity field can be derived from these data alone to wavelengths greater than about 500 to 600 km (Nerem et al., 1994). Over the oceans, satellite altimetry yields measurements of the geoid with horizontal resolutions of tens of kilometers (Rapp and Pavlis, 1990), although with insufficiently uniform accuracy at these short wavelengths for many geophysical applications. Increasingly accurate measurements from satellite altimetry (less than 10 cm for the current TOPEX/POSEIDON satellite) characterize the instantaneous sea-surface height, which can deviate from the geoid by as much as 1 to 2 m. Therefore, satellite altimetry is not the final answer to better marine gravity models.

Gravity measurements made from a moving vehicle, such as an aircraft or ship, are contaminated by motion-induced and inertial accelerations due to the rotating coordinate frame, even when mounted on a stabilized platform. These accelerations must be accurately determined and removed from the measurements to recover meaningful gravity anomalies. The amplitude of the vertical acceleration can be much greater than that of the geologically and geophysically significant anomalies. For example, a sedimentary basin may have a gravity anomaly of 125 milligal (mGal = $10^{-5}ms^{-2}$), whereas the typical vertical accelerations of an aircraft are approximately 20,000 mGal. Also, the eötvös correction (Coriolis and centrifugal effects), being a function of flight direction and velocity, can reach amplitudes of over 1,000 mGal and must be calculated for gravity measurements.

On ships, the requirements on position and velocity accuracy are less demanding because of the lower speeds and the somewhat predictable average vertical position. These less stringent positioning requirements have permitted widespread use of marine gravimetry through the use of satellite positioning and radio navigation systems, such as Loran-C. Early airborne gravimeter measurements also used Loran-C, as well as radar, laser, and barometric altimetry. These techniques, however, were largely imprecise and regionally limited. It was not until the late 1980s that precise kinematic positioning with GPS was shown to meet the rigorous positioning requirements for airborne gravity measurements (Brozena et al., 1989). GPS positioning, which is easily made on global scales, has led to the broader application of airborne gravity surveys, both in scientific research and in commercial exploration. Numerous basic and applied research groups in the United States, Canada, Switzerland, and Germany are actively pursuing both proof of concept and operational airborne gravity capabilities that are now possible with GPS.

A recent gravimetric survey of the subcontinent of Greenland illustrates the powerful new capabilities of GPS-navigated and-positioned airborne gravity surveys (Brozena et al., 1992). Because thick ice covers more than 96 percent of the surface of Greenland, the geology of the subcontinent is poorly known. Scientists flew a detailed gravity survey of the land and coastal waters of Greenland using a P-3 Orion aircraft. The aircraft was equipped with a LaCoste-Romberg air-sea gravimeter mounted on a three-axis stabilized platform, a GPS receiver, and pressure and radar altimeters. Over the course of two years, the scientists were able to map the entire subcontinent at a line spacing of 50 to 100 km with an accuracy of 4 to 6 mGal. The resulting data suggest the location of major sutures between the northern and southern regions (see Figures 1.4(a) and 1.4(b)). This survey would have been impossible without the precise positioning capabilities of differential GPS.

MEASURING THE EARTH'S SURFACE TOPOGRAPHY

Measurements of surface elevations are fundamental to many applications of geology, geophysics, hydrology, terrestrial ecology, geomorphology, glaciology, and atmospheric physics (Table 1.3). For example, the topography of the polar ice caps and mountain glaciers is important because it is a direct measure of ice flow dynamics and is closely linked to global climate and sea level change. Existing topographic data in "well mapped" North America, western Europe, and Australia, however, are inadequate in terms of accuracy and consistency to support most types of scientific research (Topographic Science Working Group, 1988). For example, in high-relief terrain, the vertical accuracy of available data may be worse than 30 m, too poor by at least an order of magnitude to support many applications. Less accessible regions of the Earth, including large areas of Africa, Asia, South America, and Antarctica, do not have even this level of topographic coverage (Figure 1.5).

Topography has traditionally been measured by triangulation and levelling methods. This land-based effort required skilled crews to measure accurately distances and elevation changes. Over the past several decades, ground determinations of topography have been supplemented by airborne photogrammetry, which provides the potential to map the Earth's surface to a decimeter. Photogrammetric techniques are adequate to meet topographic mapping standards for scales of 1:10,000 or less, but they are not sufficiently precise to monitor dynamic changes in the Earth's surface. The signals of interest to earth scientists are typically on the order of

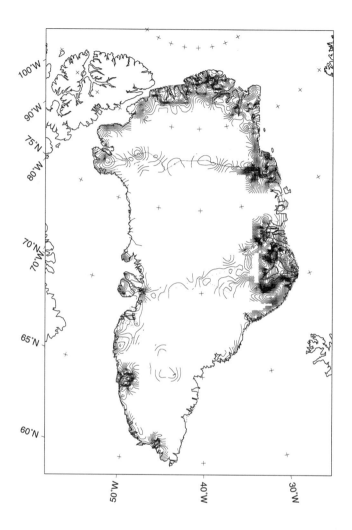

FIGURE 1.4(a) Terrestrial gravity data distribution over Greenland as of 1992, reflecting more than 30 years of ground survey efforts. The contour interval shown is 10 mGal. Contours are masked if no data point exists within 40 km. (Figure courtesy of J. Brozena, Naval Research Laboratory).

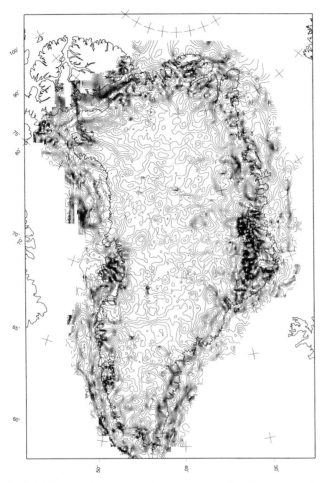

FIGURE 1.4(b)   Airborne gravity data distribution from the Greenland
Aerogeophysics Project, flown during the summers of 1991 and 1992 by the Naval
Research Laboratory and the Naval Oceanographic Office in collaboration with
NOAA and the Danish National Survey and Cadastre.   Approximately 200,000
line-km of tracks were flown over a period of 4 months, covering the entire
subcontinent of Greenland ($>2,100,000$ km²).   Interferometric-mode GPS
provided aircraft positioning, and magnetics and surface topography were mapped
simultaneously with gravity.   Contour interval is 10 mGal.   The airborne data
were combined with historical terrestrial and shipboard data, except the traverse
near 77°N, which proved to be in error by more than 30 mGal over the ice cap.
(Figure courtesy of J. Brozena, Naval Research Laboratory).

TABLE 1.3　Topographic Accuracy Needed for Earth Science Applications (Compiled from Burke and Dixon, 1988; Coolfont, 1991; Dixon et al., 1989; Fletcher and Hallet, 1983; Garvin and Williams, 1990; Qidong et al., 1984; Simkin et al., 1981; Telford et al., 1990).

Feature	Vertical Accuracy (m)	Horizontal Accuracy (m)	Repeat Interval (yr)
Geology/Geophysics			
Plate Boundaries and Intraplate Deformation			
Large-scale structures	10	1,000	-
Rifts	20	2,000	-
Diffuse extension	10	1,000	-
Mountains	10	500	-
Land Geology/Fault Zone Tectonics			
Mapping	4	30	-
Surface structure	1	10	-
Neotectonics	2	100	5-20
Volcanology			
Flow and ash volumes	0.5-3	30-100	3
Volcano morphology	2-10	30-500	-
Volcano dynamics	0.15-1	30-100	1
Marine Geology/Geophysics			
Topography prediction from geoid	0.1	3,000	-
Gravity/Magnetics			
Admittances—gravity	10	10,000	-
Terrain correction—gravity	1	200	-
Satellite gravity	3	1,000	-
Magnetics	10	200	-
Polar Science			
Basic Inventory			
Large-scale features (ice domes, divides, streams, ice shelves, drainage basins)	10	500	5-10
Medium-scale features (flow lines, undulations, crevasses, rifts)	1	100-500	5-10
Mass Balance and Dynamics			
Accumulation	0.1	500	5
Ice dynamics (gradients, flow features)	0.1-0.5	100-500	1-5
Ablation (grounding lines, ice shelf margins, rifts, crevasse fields, icebergs)	0.1-0.5	100-500	1-5

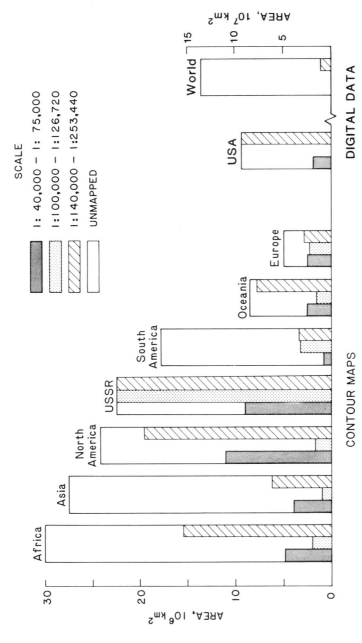

FIGURE 1.5 Global availability of topographic data at a variety of scales derived from contour and digital maps. (Figure from United National Development Project).

millimeters to tens of centimeters for postseismic rebound, for example; millimeters for stream erosion rates; and tens of meters for mass wasting processes. Precise positioning is required for these applications.

With the implementation of geodetic networks supported by land-based GPS crews, points can now be located on the order of millimeters. Recent studies have documented plate motions and have raised important questions about the nonrigid nature of the Earth's deformation along plate boundaries. A drawback of this technique, however, is that the geodetic networks provide only point measurements, and knowledge of the entire three-dimensional surface is required if earth processes are to be understood. The recent integration of airborne mapping techniques with precise positioning has demonstrated the ability of this approach to detect dynamic processes.

A striking example of the success of this approach was the detection of a subtle but significant depression in the West Antarctic ice sheet (Blankenship et al., 1993). Although the region had been photographed, surveyed for topography, and imaged by satellites, these earlier efforts failed to identify a depression in the ice surface that was 50 m deep, 6 km wide, and 12 km long (Figure 1.6). An aircraft survey with 5-km spacing equipped with a laser profiling system integrated with GPS positioning revealed that the depression was associated with active volcanism beneath the ice sheet. This discovery has important ramifications for the dynamics of ice sheet collapse, and yet the depression never would have been resolved without precise positioning.

VERY HIGH RESOLUTION STUDIES WITH AIRBORNE TECHNIQUES

Land-based gravity measurements have long been used to identify the salt domes and other structural traps for oil along the Gulf Coast and the large mineral deposits in the western United States. As the easily identified reservoirs and mineral deposits become depleted, however, exploration industries will require increasingly detailed information on the subsurface of the Earth. High-resolution mapping of the subsurface is also rapidly developing as a requirement for the characterization and cleanup

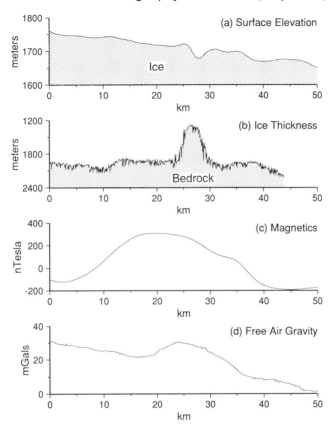

FIGURE 1.6 Evidence for active volcanism beneath the West Antarctic Ice Sheet from precise surface altimetry measurements and other airborne geophysical observations. The measurements were made as part of a major study of the stability mechanisms of the West Antarctic ice sheet and were collected along a north-south profile. (a) Surface elevations from a GPS-positioned laser altimeter reveals an anomalous depression in the ice surface located at 28 km. (b) Depth to bedrock (ice thickness) from ice penetrating radar observations. The prominent feature at 26.5 km is centered in a shallow rimmed caldera. (c) Total magnetic field observations reveal a large anomaly between 0 and 40 km that is strongly correlated with the caldera and central edifice. (d) Free air gravity anomaly of 7 mGal is associated with the central edifice at 24 km. (Figure modified from Blankenship et al., 1993).

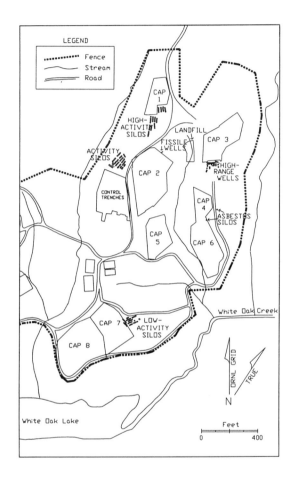

FIGURE 1.7(a) Map showing waste sites and buildings in the survey area, Oak Ridge National Laboratory. (Figure modified from Doll et al., 1993).

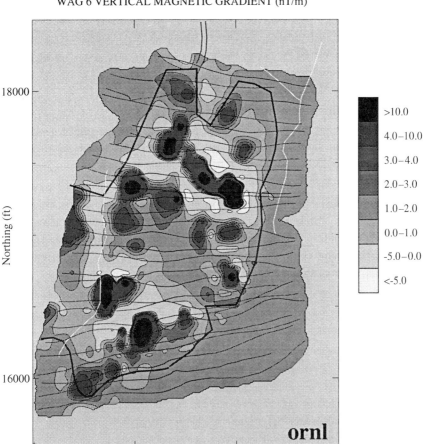

FIGURE 1.7(b) Short-wavelength magnetic anomalies (vertical component) measured from a GPS-navigated helicopter are associated with high- and low-activity silos, surface pipes (labelled "high range wells"), high- and low-activity materials disposal trenches (capped areas 1, 2, 4, 7, and Control Trenches), biological disposal trenches (capped areas 5 and 8), and asbestos disposal trenches (capped area 6). (Figure modified from Doll et al., 1993).

of Superfund and other waste sites (e.g., Labson et al., 1993; Phillips, 1993). Measurements of these waste sites often pose health hazards to ground crews, so there is great interest in the development of remote tools. These high-resolution applications can be addressed by airborne geophysical technology and precise positioning.

The ability of airborne techniques to characterize waste sites safely and rapidly was recently demonstrated at Oak Ridge National Laboratory (Doll et al., 1993). An innovative approach was developed utilizing a helicopter equipped with GPS and airborne electromagnetic, magnetic, and radiation sensors. A reconnaissance survey was flown at an altitude of 30 m with a line spacing of 45 m to detect large environmental targets and to identify the faults and fractures that could influence contaminant migration. Preliminary results showed that electromagnetic and magnetic anomalies corresponded to radioactive, biological, or asbestos waste sites, and ferrous objects, such as drums, silos, trenches, and well casings (see Figures 1.7(a) and 1.7(b)). Simultaneously, these measurements provided constraints on the regional hydrologic systems and geologic framework that partly control the extent of the hazardous material.

2

SCIENTIFIC FRAMEWORK

Airborne techniques are powerful research tools for a broad variety of applications, from resource assessment to exploratory geophysics to vegetation analysis. This chapter examines the enhanced scientific capabilities that precise GPS positioning provides to airborne geophysics. Three scientific techniques have been dramatically enhanced with precise positioning:

1. accurate regional airborne mapping of topography;
2. regional airborne gravity mapping; and
3. high-resolution surveys of topography, gravity, magnetic, and electromagnetic properties.

Application of these techniques will result in numerous advances in basic and applied research. The scientific objectives that are now possible with precise positioning and airborne geophysics can be categorized in the following areas:

1. interdisciplinary earth science studies;
2. continental geodynamics;
3. economic, environmental, and nuclear nonproliferation issues;
4. geodetic studies; and
5. global change monitoring.

These objectives were given high priority in the report *Solid-Earth Science and Society* (NRC, 1993).

INTERDISCIPLINARY EARTH SCIENCE STUDIES
AND AIRBORNE GEOPHYSICS

In recent years it has become clear that physical processes that had been studied in isolation are in reality linked. For example, geomorphic studies must incorporate an understanding of regional geology, as well as of climatic, tectonic, and hydrologic conditions. Similarly, studies of ice sheet stability must incorporate not only detailed understanding of ice properties, but also knowledge of the nature of surrounding oceans and underlying geology. Any facet of these problems may be studied in isolation, but complete understanding of the system will only come from examining the contribution of all of its components.

The research community, recognizing the importance of interdisciplinary studies and the need to share resources, has fostered the creation of cooperative research laboratories and oceanographic research vessels. The ocean science community in particular has witnessed fruitful cooperation between oceanographers studying hydrothermal mixing, marine biologists studying vent communities, and geophysicists studying magmatic processes. Airborne research platforms have a similar potential for fostering such research in a cost-effective way. As discussed below, interdisciplinary research topics that can be addressed by airborne geophysics integrated with precise positioning include the following:

- ice dynamics and sea level rise;
- erosional processes and landform development; and
- hydrologic cycle.

Ice Dynamics and Sea Level Rise

The melting or collapse of large ice sheets is believed to be responsible for rapid sea level rises documented in the geologic record. Our understanding of the collapse of ancient ice sheets is largely based on studies of the modern ice sheets that cover much of Greenland and Antarctica. The water stored in these ice sheets has vast potential for raising sea level globally. For example, the West Antarctic ice sheet, were it to melt completely, is capable of triggering a global sea level rise of 6 m (Mercer,

1978). When an ice sheet collapses, the ice is delivered to the oceans where it floats, causing a rise in sea level. Large rivers of ice (60 to 100 km wide) form; they transport the ice from the continent's interior to the ocean at speeds of up to 750 m/year. These ice rivers are critically dependent upon the presence of an oversaturated till at the ice-rock interface, which in turn is controlled by the underlying geology. Thus, understanding the ice dynamics and its potential contribution to global sea level rise is a problem requiring collaboration between glaciologists, oceanographers, and geologists.

Airborne geophysics can provide important constraints on this interdisciplinary problem. Accurate topographic measurements of 1-m resolution on the ice surface and 5-m resolution at its base are necessary to determine the stress state of an ice sheet. Simultaneous measurements of airborne gravity (accurate to 3 mGal) and magnetics (accurate to 1 nanotesla [nT]) can be used to locate volcanic sources of excess heat and the source and distribution of lubricating sediments. Incorporation of airborne electromagnetic methods would detect the thermal anomaly beneath the ice sheet.

Erosion Processes and Landform Evolution

Erosion is a complex process that depends on the nature of the bedrock being eroded, landform relief, steepness of slopes, climatic conditions, vegetation type and distribution, and hydrologic flow patterns. The relative importance of these variables has been difficult to determine because the variables are linked and because the rates and physical properties are difficult to measure. Yet, it is important to better quantify and understand erosion processes because of their importance in building construction, bridge and road stability, waste site evaluation, and even oil exploration.

An airborne research platform provides an ideal facility to study erosion processes. Such a study would include accurate measurements of topography (accurate to better than 1 m) to quantify erosion rates, high-resolution airborne magnetic and gravity measurements to define the nature of the bedrock accurately, and remote sensing images to determine the effect of vegetation. Repeated airborne surveys would show how landforms evolve with time.

Hydrologic Cycle

As the world's population expands, the distribution and allocation of water resources will become increasingly important. Water disputes in arid regions such as the western United States, the Middle East, and Africa illustrate the importance of water to society and politics. In Texas there are concerns about freshwater levels in the aquifers, as well as deterioration in water quality due to saltwater intrusion and oil contamination in heavily pumped regions. To ensure a clean water supply, it is necessary to adopt an approach that will determine the sources of contaminants and the processes that control their dispersion through the aquifer system.

Airborne platforms can provide additional information to constrain groundwater hydrology problems. Precisely navigated, high-resolution airborne magnetics can identify abandoned well heads and the source of fluid contaminants, and electromagnetic techniques can show the distribution of saline water. Airborne gravity and magnetic measurements, which are accurate to 1 mGal and 0.01 nT, respectively, can also be used to locate structures such as dikes, faults, and fractures that control the accumulation and movement of groundwater. For example, an airborne magnetic survey over Western Australia clearly delineates the subsurface trace of igneous dikes and fractures (Figure 2.1). In this region of Australia, the groundwater tends to dam up against the dikes and mix with salt-concentrated soil, causing agricultural losses. Similar magnetic surveys have also been flown to map the subsurface extent of igneous units that influence the flow of groundwater away from a weapons testing area (Grauch et al., 1993).

Airborne magnetic methods, combined with other geophysical data, can also be used to determine the thickness of strata above the magnetic basement. This information can be used to help assess the groundwater potential of a region (e.g., Babu et al., 1991). Finally, airborne topography measurements with accuracies of better than 10 cm can indicate the rate and form of subsidence caused by fluid withdrawal.

FIGURE 2.1 Airborne magnetic image over the Yornaning Catchment area, Western Australia. The linear features are dolerite dikes, which dam the groundwater resulting in surface salinity outbreaks causing degradation of farming land. (Figure courtesy of B. Larson, World Geoscience Corporation).

CONTINENTAL GEODYNAMICS
AND AIRBORNE GEOPHYSICS

Plate tectonics provides a useful template for studying geodynamic processes both on continents and within ocean basins. Studies of seafloor magnetic anomalies, fracture zones, and earthquake focal mechanisms have revealed a record of the relative motions of the Earth's torsionally rigid plates over tens of millions of years. Our understanding of continental geodynamic processes, however, lags greatly behind our knowledge of oceanic lithospheric processes for the following reasons:

1. In many places the continental lithosphere is deforming in complex ways and not simply by the rigid body rotation that characterizes the oceanic lithosphere;

2. Continental lithosphere is difficult to subduct and preserves a much longer and more complex history of accretion and deformation than does oceanic lithosphere; and

3. The rheological profile of continental lithosphere, principally its continental crust, allows for significant subsurface decoupling, making it more difficult to relate upper and lower lithospheric processes.

Airborne geophysics has the potential to advance our understanding of continental dynamics, particularly in the following areas:

- Active tectonics: complex deformation and nonrigid behavior near the Earth's surface;
- Volcanology: processes of inflation, eruption, and degradation of volcanic features; and
- Regional geodynamics: evolution and rheologic properties of the continental lithosphere.

The contributions that airborne geophysics can make to these research areas are described below.

Active Tectonics

Plate tectonics provides a framework for studying the evolution of geologic and geophysical processes through time. On global scales, the lithospheric plates are rigid and the plate boundaries are sharply defined within oceanic lithosphere but are broad and poorly defined within continental lithosphere. On regional scales, plate boundaries in both continental and oceanic settings are complex and can best be studied in regions of young and active tectonics. Within oceanic lithosphere, the structural complexity along plate boundaries may be tens of kilometers wide; within continental lithosphere, it may be 2,000 to 3,000 km wide (Argand, 1924; Molnar and Tapponnier, 1975). While deformation at plate boundaries is dominated by structures related to overall plate motion, it characteristically consists of many small fragments of crust that respond individually to global plate motions and that may be partially or wholly decoupled from deeper parts of the lithosphere (Figures 2.2(a) and 2.2(b)). For example, the convergent plate boundary between the subcontinent of India and Eurasia is marked by a zone of deformation more than 2,000 km wide extending from northern India into Mongolia (Baljinnyam et al., 1993). Not only is this deformation zone wide, but like all plate boundaries within continental crust, it contains a complex but integrated system of shortening, strike-slip, and extensional structures, even though the plates are converging. Regions undergoing principally extension may also be complex. Although extensional regions in East Africa are characterized by relatively narrow zones of deformation, extension of the 600-km-wide Basin and Range Province is diffuse. Moreover, while the region is dominated by extensional structures, strike-slip and compressional features are also found. Finally, transform boundaries, where the plates move horizontally past one another, such as along the San Andreas Fault in southern California, are often represented as thin lines on maps. Patterns of structures and seismicity, such as have occurred during the 1994 Northridge earthquake and other earthquakes in the southern California region, however, indicate that the transform boundary is characterized by a complex system of compressional, extensional, and strike-slip structures.

Much of the Earth's seismicity and volcanic activity is associated with these complex plate boundaries. At the same time, much of the Earth's population resides in cities likely to be affected by these natural hazards.

36

FIGURE 2.2(a) Topography and Bouguer gravity profiles (dots) across the Tibet Plateau oriented perpendicular to the Himalayan thrust front along the 3 transects shown in the map panel (b). The solid lines superimposed on the Bouguer gravity show the predicted gravity assuming Airy isostatic compensation of the plateau. While overall the plateau appears to be isostatically compensated, the fact that the predicted Bouguer gravity is less negative than the observed at the edges of the profile demonstrates that the Indian plate to the south (left) and the China plate to the north (right) are flexing as rigid plates. Furthermore, the 100-km-wavelength undulations in the observed Bouguer gravity are predicted by no model of Airy or elastic-plate compensation, but can be explained by compressional folding of a rheologically layered plate with a weak, ductile lower crust. (Figure from Jin et al., 1994).

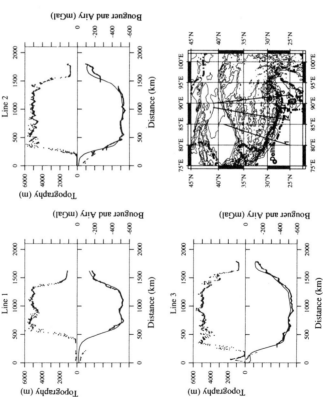

37

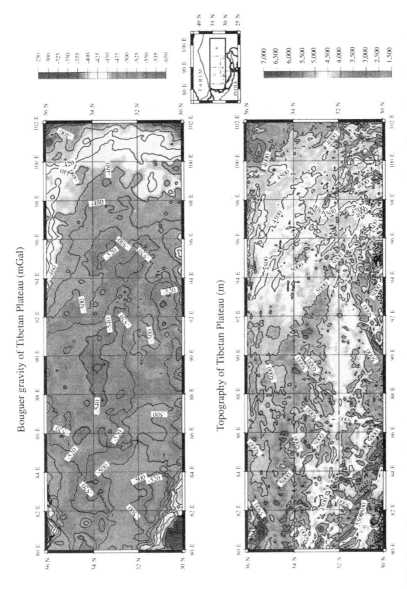

FIGURE 2.2(b) Map of Tibet Plateau showing topography and Bouguer gravity. (Figure from Jin et al., 1994).

To improve our understanding of the nature of the plate boundaries and the mechanisms that create them, further studies are needed for the following purposes:

- to determine the spatial and temporal strain distribution and the structures that partition that strain;
- to document the relationship between the regional geology and the deformation within the boundary; and
- to relate the surface expression of deformation within the plate boundary to the motion and behavior of the underlying mantle in order to elucidate the driving mechanisms for the deformation.

To address these needs, research strategies need to include ground-based techniques, such as geologic mapping, paleomagnetic analysis, studies of seismicity and paleoseismicity, and geodetic measurements, as well as airborne geophysical methods. In particular, airborne geodetic techniques could complement these other research strategies. For example, accuracies of 1 to 10 millimeters per year are needed to map the strain across plate boundaries. The majority of these measurements are currently obtained either by periodic geodetic measurements or by permanently operating GPS arrays. Both of these approaches provide detailed insight into the motion of individual points along complex fault systems, but individual sites may be contaminated by local noise or may not be representative of the area. Airborne swath mapping techniques, such as SAR or scanning laser systems, would provide sufficiently accurate measurements of the Earth's surface between the geodetic monuments.

Documenting the structure of the crustal blocks that make up a deformation zone also requires detailed understanding of the local geology. Geologic mapping integrated with multichannel seismic reflection and refraction studies can be used to determine the local structure, but these studies can be limited by poorly exposed outcrops or by high cost. Precise airborne geophysical techniques, particularly airborne magnetics and gravity, can be used to help map the local geology efficiently and economically. These methods are capable of producing the required accuracies of several picotesla and 1 mGal over wavelengths of several hundred meters. Airborne techniques also offer the advantage of mapping larger areas than can be conveniently mapped using surface methods. This allows the high-resolution studies along plate boundaries to be placed in a regional structural and tectonic framework. Such a framework is crucial for merging local observations with global plate motions and with the structure of the underlying mantle.

Volcanology

Volcanic eruptions destroy life, cause millions of dollars worth of property damage, and may even affect global climate patterns. Despite these adverse impacts on society, volcanic processes are still poorly understood. Airborne methods, particularly accurate topographic mapping, airborne gravity, magnetics, and gradiometry, can be used as follows:

- to develop models of volcano growth and degradation;
- to help determine the volume of erupted material; and
- to quantify inflation and eruption processes.

Models of terrestrial volcanic productivity, growth, and degradation depend on measurements of a number of parameters, including the morphometry of the volcano (area, altitude, volume, flank slope, profile, crater size), the composition, eruptive style, explosivity, and tectonic setting. Morphometric characteristics of individual volcanos can be derived from digital topographic data, but systematic studies of suites of volcanic landforms at scales larger than ~ 1 km are needed to provide model input (e.g., Pike and Clow, 1981). Recent work by Garvin and Williams (1990) has demonstrated that Digital Elevation Model (DEM) data, with ~ 100-m horizontal resolution and 10- to 20-m vertical resolution, are suitable for characterizing 1- to 20-km-scale volcanos.

Accurate topographic data are required to estimate eruption volumes. The volume of volcanic deposits, however, is difficult to measure because of uncertainties in the boundaries of the deposit and in the topography of the pre-eruption surface. High-accuracy digital topography, with 0.5- to 3.0-m vertical resolution and 30- to 100-m horizontal resolution, is required to resolve deposit boundaries.

To model volcano inflation, deformation, and pre-eruptive surface modifications, it is necessary to obtain sub-meter geodetic measurements (vertical) on topographic length scales of 30 to 100 m on historically active volcanos. The integration of airborne measurements of accurate topography with high-resolution gravity or gravity gradiometry would greatly enhance our understanding of volcano dynamics.

Regional Studies

Regional studies provide the framework for understanding the dynamics of the continental lithosphere. Surface gravity and regional topography data play a crucial role in constraining the mechanical and thermal properties of the lithosphere. For example, dense gravity surveys, with accuracies of 2 to 3 mGal at resolutions better than 10 km, integrated with topography data can characterize crustal thinning within continental rift zones. Detailed gravity data have also revealed the complex state of isostatic compensation, and, thus rheology of mountain chains formed by continental collision, although there are still unresolved questions on what kinds of compensation mechanisms are associated with the different types, locations, and histories of plate collision. Finally, high-resolution gravity data over sedimentary basins are needed for an understanding of the thermal and mechanical driving forces that cause basin subsidence. For example, gravity measurements have been used to interpret the internal structure of the Dead Sea basin and model its evolution (e.g., Ten Brink et al., 1993). Marine gravity measurements show that the Dead Sea graben is isostatically uncompensated, consistent with its formation by pure mechanical stretching within the crust.

In summary, gravity data with accuracies of 1 to 10 mGal and resolutions of 100 to 200 km or better provide a crucial component in the understanding of the dynamics of the Earth's lithosphere (NASA, 1987). Accurate topography data, however, are necessary for interpreting the gravity data and for studying internal density variations and elastic strength. A standard analysis method is to compute a linear transfer function (admittance function) that best maps topography into gravity for a particular tectonic region. The assumption is that the topography acts as a load on the elastic lithosphere, which responds linearly and in phase with the load, so that the relationship between topography and gravity is uniquely specified by a linear admittance function. The differences between gravity-topography admittance functions become apparent for wavelengths greater than about 20 km. For this application, the topography should reflect the average elevation in a 10 × 10 km grid cell. When working with 1-mGal accuracy gravity data, the areal averaged topography needs to be accurate to about 10 m.

HIGH-RESOLUTION APPLICATIONS AND AIRBORNE GEOPHYSICS

Resource Exploration

Exploration for natural resources has long depended on geophysical remote sensing techniques. With the advent of high-resolution gravity and magnetic mapping from aircraft, there is an opportunity to better meet the needs of the mineral and petroleum industries. Further advances in the resolution and accuracy of these techniques will serve to broaden their application by industry.

A major focus for hydrocarbon exploration is to establish which sedimentary basins will become major producers in the future (NRC, 1993). Although seismic techniques are the dominant exploration tool, high-resolution gravity and magnetic measurements collected rapidly and efficiently from airborne platforms will have increasing appeal in the petroleum industry. GPS and new sensor technologies now allow the following features to be identified in sedimentary basins using airborne methods:

- structures beneath thick sheets of salt that cannot be easily imaged with seismic techniques because of the velocity structure of salt;
- faults which may play an important role in oil migration; and
- structures within the sedimentary column that may trap oil.

Structures that trap oil and gas within sedimentary basins can be located with gravity and magnetic surveys. For example, salt sheets and diapirs produce gravity lows and anticlines produce gravity highs. The gravity signature of these structures is on the order of 1 to 10 mGal with wavelengths of 1 to 10 km. Airborne gravimetry is ideally suited for regional gravity mapping at these wavelengths and levels of accuracy and can be used for rapid exploration of hydrocarbon prospects (e.g., Gumert, 1992). In addition, high-resolution gravity gradiometry measurements can help resolve the nature of the basin strata and high-resolution magnetics (sub-nT) can provide critical information on the form of faults and the detailed structure of sedimentary rocks.

The minerals industry typically focuses on even smaller targets than those that interest the petroleum industry. The location of large ore deposits in the shallow crust can be discerned from gravity surveys, but their small size (10 to 100 m) and low amplitude (< 1 mGal) make them difficult to map in detail. The development of high-resolution airborne techniques that can map small features was partly driven by the minerals industry in Canada and Australia. These countries routinely use high-resolution magnetics and gravity and current studies are exploring the use of gravity gradiometry where a resolution of 2 eötvös (10^{-9} s^{-2}) could provide the level of detail necessary to a successful exploration project.

Nuclear Verification

With the conclusion of the Cold War and rapid changes in the former Soviet Union, issues of nuclear verification and nonproliferation have taken a different form. The prime threat is no longer perceived to be Russia or Ukraine but the development of nuclear capabilities by less developed countries. The present strategy involves a tiered detection and monitoring effort that incorporates global, regional, and local seismic networks and on-site inspections (U.S. Congress, Office of Technology Assessment, 1988). Airborne techniques, particularly swath mapping techniques that can cover large areas rapidly and remotely, should be part of the strategy for identifying the location of subsurface tests and for monitoring postshot subsidence. Subsurface test sites could be located by identifying excavation areas and characteristic surficial expressions of an explosion site. In the United States, underground nuclear tests typically create large, circular surface depressions (Figure 2.3) that are 2 to 70 m deep, and 20 to 130 m in diameter (e.g., Houser, 1970). The subsurface cavities could be identified with airborne gravity, gradiometry, or electromagnetic measurements; the surface expression of a test site could be determined with accurate topographic mapping. Repeated topographic mapping of a region could also identify postshot subsidence. Work at U.S. test sites has documented continuing displacements on the order of tens of centimeters in the 18 months following an event. Airborne observations

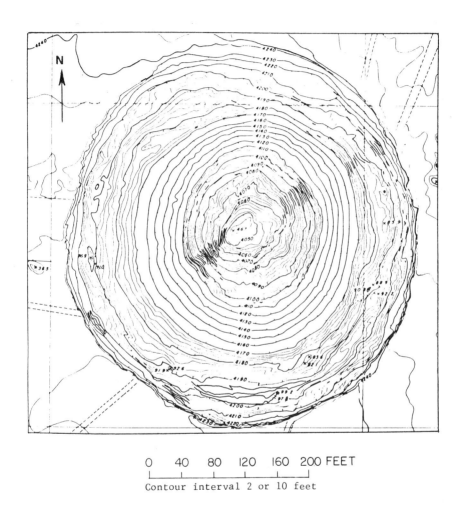

0   40   80   120   160   200 FEET

Contour interval 2 or 10 feet

FIGURE 2.3 Topographic map of a large, symmetrical sinkhole produced by an underground nuclear test, northeastern Yucca Flat. (Figure from Houser, 1970).

could supplement seismic detection of the occurrence and timing of a nuclear test.

GEODESY AND AIRBORNE GEOPHYSICS

One of the principal objectives of geodesy is to determine the shape of the Earth's surface by determining land and sea heights relative to a reference surface. The traditional reference is idealized mean sea level, or the geoid. The geoid is an equipotential surface in the Earth's gravity field; it can be determined from surface gravimetry data, either as a surface integral of gravity or as a line integral of the deflection of the vertical (the slope of the geoid).

On land, the geoid is the principal reference for heights; it is approximately determined from mean sea level observations at coastal tide gauge stations. Discrepancies of up to a meter between tide gauge stations on different oceans, or even along longer coastlines of the same ocean, indicate that mean sea level does not coincide exactly with the geoid (a unique surface). Comparison of vertical reference systems (vertical datums) of different countries, therefore, requires a consistent global geoid model determined from gravity data (Rummel and Teunissen, 1988). As a rough rule of thumb, if the geoid is resolved over a distance of x kilometers, the error introduced by neglecting the unresolved part is $x/300$ meters. In order to connect vertical datums to 10 cm or to obtain a consistent reference for larger vertical networks at this level, the geoid must be resolved to a horizontal spatial scale of at least 30 km.

Heights above mean sea level (orthometric heights) are traditionally determined using spirit leveling, which is accurate to a few millimeters over several kilometers of leveling line length. A less labor-intensive procedure for determining orthometric heights is based on satellite ranging using GPS. With an appropriate number of satellites and good geometry between the satellites and the ground station, it is possible to obtain heights above the geocenter, or equivalently, above a reference ellipsoid defined in the GPS satellite coordinate system. With this method, absolute heights can be determined to an accuracy of a meter or better, while relative heights determined by differential techniques are accurate to centimeters or millimeters (depending on line length). To determine the corresponding orthometric heights, the geoid height must be obtained with commensurate accuracy. Some tests indicate that the current relative geoid accuracy in the United States ranges from 1 cm/10 km, to 10 cm/100 km, to several decimeters over longer distances (up to 1 ppm) (Milbert, 1991). For certain applications, such as photogrammetric topographic mapping, a uniform accuracy and resolution of about 2 cm/25 km is now needed in view of the potential of kinematic GPS aircraft positioning at the 2- to 3-cm level.

In summary, there continues to be a purely geodetic need for high-resolution global gravity to determine the geoid and to correlate the various vertical datums of the world. The ultimate goal is to provide a uniformly accurate and consistent vertical reference for navigation and positioning. The promise of relatively inexpensive, yet very accurate, height determinations using GPS drives a need to know the geoid to comparable accuracy and resolution. Airborne gravimetry offers a cost-effective means to map the geoid, particularly in geographically remote areas.

GLOBAL CHANGE MONITORING
AND AIRBORNE GEOPHYSICS

Monitoring Ice Sheets and Mountain Glaciers

Melting ice has been identified as the major cause of sea level rise. The majority of ice that has the potential to raise sea level is held in the East and West Antarctic ice sheets and the Greenland ice cap, but the mountain glaciers that cover much of Alaska and Patagonia may also contribute to sea level rise (Meier, 1984; Thomas, 1991). The volume of ice held in these reservoirs and its change through time are poorly understood. Existing estimates are derived from historical records and from space-based measurements. A new generation of space-based radars and lasers (e.g., ERS-1, ERS-2, RADARSAT, JERS-1, EOS-ALT) will provide topographic measurements of large areas of the ice sheets (U.S. Congress, Office of Technology Assessment, 1993). Current sensors are inadequate for most ice sheet applications (see Table 1-3), but by the end of the 1990s, higher-precision lasers and radars will be in orbit. Airborne techniques are necessary to establish a baseline now and can be used to systematically map the highest-latitude regions of the ice sheets not covered by the space missions. Moreover, aircraft provide a more flexible platform for more frequent systematic mapping of glaciers.

Monitoring Ocean Circulation

Global models of ocean circulation are determined in part by continuous observations of ocean currents by satellite or aircraft altimetry. Altimetry techniques measure the sea-surface height, and the difference between sea-surface height and the geoid height yields the sea-surface topography. The steady-state component of sea-surface topography is caused by deviations of the oceans from hydrostatic equilibrium (i.e., the equipotential and isobaric surfaces no longer coincide) because of variations in temperature (primarily a zonal, north-south effect) and salinity. Other regional effects include the trade winds and the effluent of major river systems. The buildup of water creates a horizontal pressure gradient that can only be sustained in the presence of currents under geostrophic equilibrium. The geostrophic currents extend to considerable depth, commensurate with their horizontal scale.

Because of the role of geostrophic currents in heat transfer and thus in global climatology, it is important to map their spatial and temporal distribution accurately (e.g., Fu, 1983). High-resolution altimetry and accurate geoid determination are needed as inputs to global change models as well as to other environmental and ecological applications, such as the dispersion of pollutants and the migration of fisheries. These applications require measurements on a broad spectrum of scales: ocean gyres are thousands of kilometers across, with sea-surface topography of up to a meter; boundary currents are hundreds of kilometers in extent, with topographies of tens of centimeters to a meter; and mesoscale eddies have scales of tens to hundreds of kilometers and topography on the order of a few decimeters. To determine their steady-state components, the geoid must be known at all of these scales to accuracies much better than a meter, and in many cases to accuracies better than 10 cm. Corresponding geoid "slopes" must be known to accuracies of about one part in ten to the seventh, yielding current velocities accurate to about 1 cm/s and requiring the equivalent in gravity accuracy of about 0.1 mGal. Airborne gravimetry fills the gap at the shorter wavelengths not measurable by satellite techniques, especially in coastal regions where ocean circulation dynamics are complex, and where, as it happens, operational and logistical support is available for aircraft geoid mapping missions.

Ecosystems

Topography exerts an important control on surface hydrology through its influence on intercepted radiation, precipitation, runoff, evaporation, snow ablation, soil moisture, and vegetation patterns. Topographic parameters also indicate the exposure of a landscape to weather and sunlight at a given latitude and thus are an indirect measure of its microclimate. Through feedback mechanisms, vegetation itself influences energy and mass fluxes, affecting not only the local environment but also the regional and global climate. Topography is therefore a key element in the study of complex ecosystems.

Quantitative hydrologic and ecosystem models require digital elevation data, but even high-resolution digital elevation data have serious deficiencies. These deficiencies are particularly evident when determining the topographic effects on solar radiation, which requires accurate slope and aspect information. A digital elevation grid is used to calculate the local gradient and angle to the horizon; the grid is combined with the solar geometry to determine the incidence angle of solar radiation on the slope. Noise in the digital elevation data of mountainous areas, which are often derived by digitizing contour lines from topographic maps, is magnified by the differencing operations used to calculate gradients. High-accuracy airborne topography measurements, with a horizontal resolution of 30 m or better and a vertical accuracy of 1 to 3 m or better, are needed to improve the solar radiation calculations described above.

Altimeters with the appropriate design and resolution can be used to assess several important vegetation parameters. In areas of low to moderate vegetation density, pulse shape analysis with laser profiling data allows estimates of the difference between the ground elevation and the canopy top. The pulse is scattered by the various vegetation layers as it penetrates, and, in principle, it is possible to discriminate the waveform of the various returns. Relative "brightness" determinations of each layer (related to projected leaf area) may permit calculations of extinction cross-section and Leaf Area Index, which are important in many ecological studies. Seasonal and longer-range variations in this index can also be observed with sufficient repeat coverage and ground calibration studies. It may also be possible to estimate canopy thickness and total biomass for determining total carbon storage. The repeated measurements and complete coverage offered by airborne radar interferometry can even yield estimates of deforestation rates. Other techniques, such as digital elevation maps, can be used to define barriers and corridors affecting species dispersal and to predict biomass, timber site quality, and burn patterns over large areas. The need for accurate slope and aspect data is especially great in semiarid, midlatitude regions where minor slope changes greatly affect local water availability, soil moisture, and vegetation cover over large areas.

Hydrology

Topography plays an important role in the distribution and flux of soil moisture, runoff generation, and discharge pattern. Digital topographic data are used to calculate watershed structure and, to some extent, runoff (e.g., Band and Wood, 1988; O'Loughlin, 1986). To model the surface and subsurface fluxes of water quantitatively, derivative quantities based upon elevation data (e.g., local gradient and upslope area) are required. Error propagation dictates that the original elevation data be of very high quality for use in the quantitative models described above. Because the data are derived from contour maps, which have large uncertainties in the vertical component, most digital data bases have proved inadequate for such calculations.

The influence of topography on terrestrial hydrologic processes is strongest in mountainous regions. High horizontal resolution is critical; the high topographic gradients and changing aspect in mountainous terrain render even 100-m data (the current resolution of many digital data bases) inadequate. Surprisingly, subtle topographic relief in regions of low slope (less than about 0.5) can also have significant hydrologic consequences, particularly for soil water dynamics and subsequent ecosystem processes. High vertical accuracy is required in these areas to identify subtle ridges or depressions that cause water flow to diverge or converge.

3

FUTURE DIRECTIONS

The previous chapter provides a framework for the scientific advances that are now possible with precisely-positioned aircraft. This chapter describes the technical foundation that must be developed if these goals are to be achieved. With the advent of new technologies for measurement, positioning, and outfitting airborne platforms, aircraft may well become the geophysical research vessels of the next decade.

MEASUREMENT AND POSITIONING TECHNOLOGY

Airborne Measurements of the Earth's Gravity Field

Airborne Gravity. With the kinematic positioning and navigation capabilities of GPS, airborne gravimetry is a viable tool for mapping the regional gravity of long- and medium-wavelength features. The resolution of this technique depends in part on the speed of the aircraft. The accuracy depends on the vibration environment and the stability of the platform. A high-speed, long-range aircraft such as the Lockheed P-3 Orion yields accuracies of 2.5 to 4.0 mGal and wavelengths of 10 to 20 km, whereas a medium-speed, fixed-wing aircraft such as a Twin Otter yields accuracies of 1.0 to 2.5 mGal and minimum recoverable wavelengths of 5 to 8 km. Short-wavelength targets can be met by using helicopters surveying at speeds of 10 to 50 knots. This approach is capable of recovering accuracies of 1 mGal and wavelengths of 1 to 2 km but is expensive for routine use. With conventional aircraft it is difficult to measure airborne gravity to accuracies of better than 1 mGal at wavelengths less than 1 km. This limitation on resolution represents a fundamental barrier to the wider application of airborne gravity by the petroleum, minerals, and hazardous waste industries, as well as by the scientific community. The primary limiting factors on airborne gravity resolution and accuracy are the sensor design, the stabilized platform, and the positioning. Most gravity sensors are based on technology developed 20 to 40 years ago. The widely used Lacoste-Romberg zero-length spring was developed in 1943 and is still considered the industry standard for marine gravity surveys. The sensor has been widely extended to airborne application because of its reliability, but its asymmetric design makes it susceptible to contamination by aircraft motion. The second most widely used sensor, the Bell Aerospace accelerometer, has a linear design and has been primarily used for military applications. Despite the elegance of the design, the sensor is extremely temperature-sensitive and is not presently capable of providing the response necessary to improve airborne gravity measurements to the sub-mGal, sub-kilometer level. There is clearly a need to integrate a new generation of gravity sensors into the current systems.

The stabilized platform commonly introduces errors in airborne gravity measurements. The platform is designed to ensure that the sensor is aligned orthogonal to the geoid so that it measures the vertical component of gravity. If this orientation is not maintained, the measured signal will include a poorly constrained fraction of the aircraft's horizontal acceleration. As with the sensors that are widely used today, stabilization technology was originally designed for marine investigations; this technology was adapted for airborne applications nearly 40 years ago. Subsequent modification of platform technologies has yielded both longer periods and digital control, which have enhanced data recovery during turns and periods of turbulence. Nevertheless, the stabilization system continues to be a major source of error and needs to be improved to take full advantage of airborne techniques.

The principal difficulty in recovering high-accuracy gravity data at medium and long wavelengths is the reduction of differential GPS data. This process requires a major investment of analysts' time following a mission to ensure that precise positions are retrieved. A very accurate knowledge of the altitude of the survey aircraft is required for two reasons. The first concerns matching the height of the measurement to a common datum, where the error of a mismatch in height is about 0.3 mGal/m. The second is more problematic, being the requirement to calculate and remove the vertical acceleration of the aircraft. Vertical positions suitably smoothed can be time-differentiated twice to obtain vertical accelerations. In addition, horizontal velocity of an aircraft must be known to about 10 cm/s or better over the smoothing interval to compute the eötvös corrections with an accuracy of less than a mGal.

Although technological advances will improve the accuracy and resolution of airborne gravity measurements, the costs must also be reduced to enable researchers to utilize this technique fully. Costs could be reduced by (1) integrating instruments and their support equipment to achieve smaller size, less weight, and lower power consumption (thus requiring smaller, less costly aircraft), (2) optimizing the productivity-to-effort ratio by fine-tuning the techniques for data processing and survey geometry analysis, and (3) streamlining the operational logistics and supporting infrastructure to increase efficiency. These measures alone would sustain airborne gravimetry as a valuable tool for earth science studies.

New Directions In Recovering The Earth's Gravity Field. Traditional ground, ship, and aircraft techniques measure the vertical component of the Earth's gravity field. New techniques, such as vector gravimetry (Schwarz et al., 1992) and gravity gradiometry (Jekeli, 1988) offer the opportunity for extracting additional information about the Earth's structure. Vector gravimetry yields the amplitude and direction of the gravity vector, whereas gravity gradiometry yields the various gradients of gravity.

Vector gravimetry is based on a triad of mutually orthogonal accelerometers that provide sensed accelerations in all three directions. That is, the gravimeter and stabilized platform are replaced by an inertial navigation system. This idea is not new. It was demonstrated for land systems in the 1970s and 1980s. It was also considered as a useful by-product, though never fully demonstrated, of the Aerial Profiling Terrain System (APTS) developed by Charles Stark Draper Laboratory in the mid-1980s. At the same time, Northrop Corporation conducted some flight tests with a stellar-inertial platform to measure deflections of the vertical (horizontal gravity components) at altitude. The principal difficulty with vector gravimetry (where the horizontal components are of primary interest) is the orientation requirement of the platform. Since the vertical component of gravity couples with the sine of the error in leveling into the horizontal component, every arcsecond of leveling error corresponds to 5 mGal of horizontal gravity error. Therefore, the natural drift of the gyros that orient the platform requires special attention; for example, Northrop used a star tracker to limit the gyro error growth. Studies indicate, however, that approximate high-pass filters may alleviate this problem without the expense of additional attitude sensors.

In an effort to circumvent the problems of separating the gravitational signal from the total acceleration environment of a moving platform, the Defense Mapping Agency (DMA) in the 1970s and 1980s undertook a unique program for airborne gravity gradiometry for regional applications. Unlike a gravimeter, an ideal gradiometer is not affected by the linear kinematic accelerations of the moving platform on which it is mounted. Instead, the gradiometer senses the curvature of the gravitational potential field. Thus, it is also more sensitive than the gravimeter is to the local density structure. An example of this technique is demonstrated in a model of a buried meandering stream (Figure 3.1). The positioning requirements for gradiometers are only about 100 m in any direction (horizontal and vertical) for the relative accuracy goal of about 1 mGal over 10 km. Initial field tests conducted in the Texas - Oklahoma area in 1987 demonstrated that DMA's Gravity Gradiometer Survey System (flown on a C-130 transport) could conduct gravity surveys in an airborne mode with a recovery accuracy of a few milligals at a resolution under 10 km in each of the three components of the gravity vector. Unfortunately, many of the data from this first test were compromised by problems with the GPS navigation and excessive platform vibrations as well as by intermittent degradation of the gradiometer instruments and/or environmental controls. The program was terminated because of its high cost but the concepts were demonstrated successfully (Jekeli, 1993).

Currently, several gravity gradiometry sensors are being developed for airborne applications. Some efforts have evolved from planned satellite gradiometer missions, including the proposed Aristoteles mission of the National Aeronautics and Space Administration (NASA) and the European Space Agency (ESA), and NASA's proposed Satellite Gravity Gradiometer Mission (University of Maryland gradiometer). Other development efforts

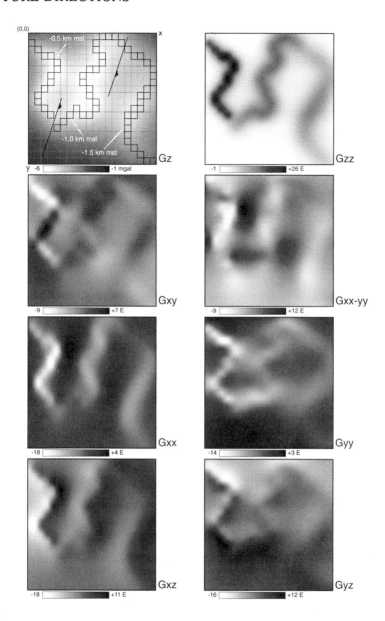

FIGURE 3.1 Model gravity data for a buried meandering stream at 0.5-km, 1.0-km, and 1.5-km burial depths. Images are 16 km across with a sampling density of 126 m. Gradient scales are in eötvös. The computed gradients (e.g., Gzz is the vertical gradient of the vertical component of the gravity vector) show greater detail than the normal gravity. (Figure from Bodard et al., 1993).

have principally been the result of the mineral industry's interest in recovering a very high resolution gravity gradient. None of these developmental systems have been subjected to the operational environment of an aircraft.

Long-term goals in airborne gravity gradiometry are to reduce the instrument size, provide adequate platform stabilization, and provide support equipment for airborne experimentation and eventual operation. Despite the technological difficulties, gravity gradiometry provides a powerful way to resolve the fine structure of the gravity field, and it should therefore be encouraged and developed as a future tool for problems in geophysical interpretation. Concurrently, airborne scalar gravimetry must be enhanced and vector gravimetry tested as the only near-term (10 to 20 years) technologies for resolving the gravity field between 10 km and several hundred kilometers.

Airborne Surface Topography Measurements

Existing Mapping Capabilities. The oceanographic community has long measured the topography of the ocean floor through remote techniques that evolved from casting weighted lines over the sides of ships, to acoustically mapping a single trace beneath a research vessel, to using the highly refined swath mapping systems. Swath systems, which recover a detailed image of the ocean floor topography in strips up to 10 km wide, are now standard equipment on state-of-the-art research vessels. The technology to map the surface of continents is undergoing similar evolution, although the development lags behind that achieved by the ocean science community. For example, current airborne topographic mapping techniques consist of rigidly mounted laser and radar ranging systems integrated with precise attitude and positioning recovery. These systems recover accuracies of about 20 cm (e.g., Garvin, 1993), which are similar to accuracies provided by early ship profiling technologies.

Swath mapping capabilities are beginning to be integrated into interferometric and stereographic techniques and into developing systems that sweep a path orthogonal to the aircraft flight path. Examples of swath systems include a laser ranging system that samples a 250- to 400-m-wide swath, and an interferometric synthetic aperture radar (SAR) system that produces swath images of topography 6 to 15 km wide with accuracies of

2 to 5 m. Photographs taken from a precisely-positioned aircraft produce an image whose accuracy and size is proportional to the aircraft altitude and the camera's focal length and film size. Examples of these four techniques are shown in Figures 3.2(a) through 3.2(d) and are described below. Table 3.1 summarizes the present range of airborne mapping techniques.

Airborne mapping techniques have several well-documented sources of error, including the following: (1) aircraft positioning; (2) vegetation, which affects most remote techniques; (3) weather conditions, which prevent the laser-ranging-based systems and the photogrammetric approach from imaging the ground surface; (4) steep surface slopes or rough terrain, which affect ranging systems; and (5) moisture content in the near surface, which affects interferometric SAR systems.

The single-ranging systems that recover topography have long been used to recover vertical positions for airborne gravity surveys over regions covered by water or sea ice. The ranges to the surface can be collected with either a radar system or a laser system. The laser systems are able to sample at a very high rate with a small footprint (≤ 1 m), but they are generally unable to penetrate cloud cover or the vegetation canopy. Recently, however, a laser altimeter with a large diameter footprint (6 to 30 m) developed at NASA's Goddard Space Flight Center has successfully imaged both the tops of the trees and the underlying ground terrain, even in densely vegetated areas (e.g., Harding et al., 1994). Radar systems have larger footprints and sample at a slower rate than laser systems, but they are generally less hindered by local weather conditions such as cloud cover. Radar ranging systems are used by commercial military and university research groups to integrate topography, gravity, and other geophysical measurements.

Simple laser ranging systems have recently been modified to sweep a path orthogonal to the aircraft motion. Such systems have been constructed and tested by NASA and commercial vendors and are capable of imaging 250- to 400-m-wide swaths with a conical scan process. The results are comparable with those of the ERS-1 satellite mission over Greenland (Thomas et al., 1992). Laser ranging systems, however, are limited by steep slopes, rough terrain, and adverse weather conditions.

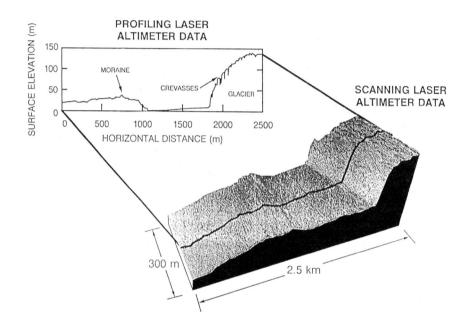

FIGURE 3.2(a) Comparison of NASA scanning laser altimeter data from the Airborne Oceanographic Lidar (AOL) sensor with independent laser altimeter profile data from the ATLAS sensor for the surging terminus of the Skeidarar-jokull outlet glacier in southern Iceland. Both data sets were acquired on September 23, 1991, as part of a NASA/U.S. Geological Survey cooperative study of landforms in Iceland. (Figure courtesy of J. Garvin, NASA/Goddard Space Flight Center and R. Williams, U.S. Geological Survey).

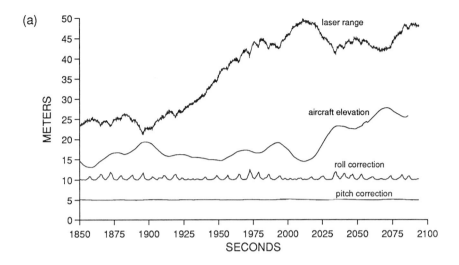

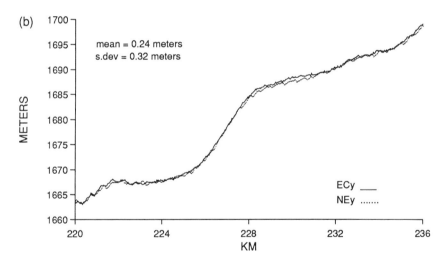

FIGURE 3.2(b) Precise surface altimetry. Top: Raw laser ranges with altitude and attitude corrections for a portion of a CASERTZ profile with an aircraft speed of about 75 m/s. The aircraft elevation is based on the differential carrier phase GPS solution and the aircraft attitude corrections are derived from the inertial navigation system. (Figure from Childers et al., 1992). Bottom: Example of repeat surface altimetry along a 16-km line. These two lines, separated horizontally by approximately 70 m, have a mean difference of 0.24 m and a standard deviation of 0.32 m. (Figure from Blankenship et al., 1992).

58

FIGURE 3.2(c) Photogrammetric images of Landers, California, taken immediately after the June 28, 1992, earthquake. The photos are stereo images of the surface rupture (horizontal line). Through the photogrammetric process, horizontal accuracies of 1:20,000 and vertical accuracies of 1:10,000 of the flight altitude are routinely obtained. (Figure from I.K. Curtis Services of Burbank. Courtesy of B. Young, Riverside County Flood Control District).

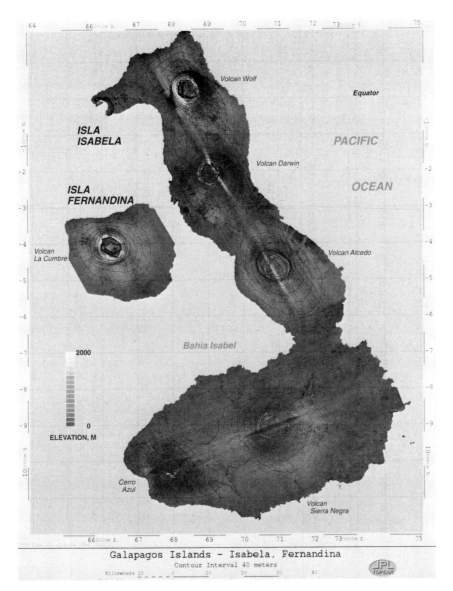

FIGURE 3.2(d) Image made from data collected for Isla Fernandina and Isla Isabela in the Galapagos archipelago using the JPL/NASA TOPSAR instruments. Over 50 40 × 12 km images were mosaicked together to obtain this DEM. Relative height accuracy is 2 to 5 m and absolute height accuracy is estimated to be 10 m. (Figure courtesy of S. Hensley, Jet Propulsion Laboratory).

TABLE 3.1 Topography Measuring Systems (Compiled from input at 1993 Workshop on Airborne Geophysics).

Systems	Swath Width	Horizontal Resolution	Vertical Resolution	Limitations	Positioning & Attitude Requirements
Profiling Laser	n/a	1-5 m	20 cm	cloud cover canopy top	0.1 m/0.2 mrad
Scanning Laser	400 m	4 m	0.1-1 m	cloud cover	0.1-1 m/<2 mrad
SAR Interferometry	5-15 km	10 m	1-3 m	all weather canopy top (single pass) partial penetration (multiple pass)	1 m/0.2 mrad 2 cm/1 mrad
Photogrammetry	scale dependent 1.5 × altitude	variable	0.25 mm × mapping scale	cloud cover canopy top	0.1-1 m/0.1 mrad

Synthetic aperture radar is a relatively new technique that allows topographic mapping to be performed in sparsely vegetated regions with a swath width of 6.4 km and an accuracy of 2 to 5 m. The prototype system is currently mounted aboard a NASA DC-8, and was developed as part of a planned low-Earth orbiting topographic mapping mission. The antennas on the DC-8 are capable of imaging with C-band (with a wavelength of 5.6 cm), P-band (68 cm), or L-band (24 cm). SAR images are used to construct an interferogram from the returns recorded at the two antennas using the C-band. This process measures the phase change due to differences in surface height, surface speckle interference, and viewing geometry between two SAR images collected at different angles. The interferogram is also used to recover topography from separate passes over the same region with either P-band or L-band. The comparison of existing digital elevation models with airborne SAR results indicate the accuracy of the airborne data is ~2 m in flat regions and ~5 m in rough topography (Evans et al., 1992; Madsen et al., 1993). The design goal of the system is 1 to 2 m.

The ERS-1 (satellite-based SAR) imaging of the Landers earthquake region is a striking example of the power of SAR to map changes in topography (Figure 3.3). The SAR interferogram demonstrates the powerful capabilities of this technique for constraining the full strain field (Massonnet et al., 1993). Single SAR images can also be used for mapping and for identifying morphologic indicators of faulting. The SAR approach is unique in that it can penetrate clouds and is sensitive to subtle changes in surface slope and roughness.

Applications of interferometric SAR either from satellites or aircraft are best suited for arid, sparsely vegetated regions because rainfall and vegetation create a temporal decorrelation between the images. The decorrelation problem may be addressed with the installation of corner reflectors and the use of ground control. These measures allow changes on the order of 5 mm to be detected (Guignard, 1992). The positioning and orientation requirements are critical, ranging from 1 m and 0.2 millirad (mrad) for single-pass C-band applications, and 2 cm and 0.1 mrad for multiple pass P- and L-band applications. High-quality navigation also becomes necessary when separate passes over a region are required.

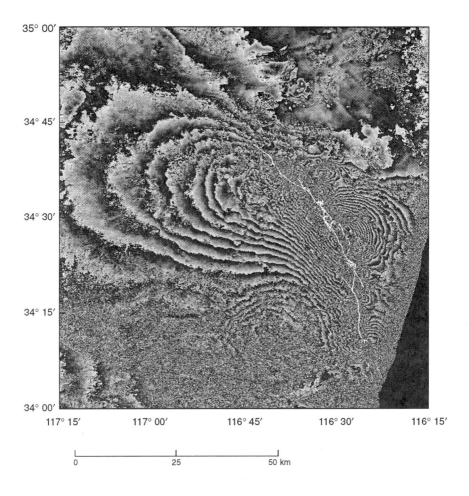

FIGURE 3.3 Detail of coseismic displacements near the June 1993 Landers earthquake determined using interferometric synthetic aperture radar techniques. Each cycle of gray corresponds to a 28-mm displacement toward the spacecraft. (Figure from Massonnet et al., 1993. Courtesy of K. Feigl, Observatoire Midi-Pyrénées, Toulouse).

Airborne photogrammetry is a well-proven technique that uses overlapping aerial photographs to compile topographic maps, for example. The concept is to geometrically relate the camera center, photograph, and ground reference system using bundles of light rays. Traditional photogrammetry relies on the use of ground control points that can then be used to relate a photograph to a three-dimensional reference system, the so-called external orientation process. With the development of precise aircraft positioning techniques, this expensive ground control can be minimized or even removed. The goal is to position the camera accurately at times of exposure so that control is acquired at flight level. Required accuracies vary according to the scale of the photography and generally range from a few centimeters for large-scale mapping to a few meters for small-scale projects. Precise aircraft orientation is also required for single-strip photography without ground control, since there is no definition of the roll component.

Recent developments in airborne mapping include the application of precise positioning and orientation to digital imaging systems (e.g., see Schwarz et al., 1993). As with photogrammetry, accuracy requirements are a function of the digital camera (i.e., spatial and spectral resolution) and flying altitude. In general, required aircraft accuracies do not exceed 0.5 m.

Future Advances in Mapping Technology. Advances in the ability to position airborne platforms precisely has led to rapid evolution of high-resolution topographic techniques. Problems still exist, however, in the following areas: (1) limited usage, which could be alleviated by integrating swath mapping and other topographic technologies with other geophysical techniques; and (2) ground imaging through vegetation, which could be improved by refining both the instruments and the algorithms.

Topographic mapping surveys are rarely flown in conjunction with other geophysical objectives. For example, scanning laser systems are frequently flown in isolation, and photogrammetric missions are flown with little in the survey aircraft except the camera and a GPS receiver. To become more effective geophysical tools, these systems must be integrated with other measurements. This would require a significant effort at reducing the size and power requirements of some systems (e.g., SAR) and ensuring that the technology is available to incorporate these mapping approaches into broad geophysical surveys.

Vegetation, particularly thick and high vegetation, provides a special challenge in airborne topographic mapping. The principal effort under way to circumvent this difficulty is the development of the small footprint laser, which is capable of imaging through holes in the forest canopy. The small footprint and full waveform recovery may lead to the capability of imaging the Earth's surface through vegetation. These and other efforts should be expanded.

KINEMATIC GPS TECHNOLOGY

GPS technology plays an important role in meeting the positioning, navigation, and orientation requirements of airborne geophysical surveys. These are all kinematic GPS applications because they involve recovering information on the location of a moving object. This section reviews the role of GPS and the present limitations on accuracy for each of these applications.

Positioning Technology

The standard approaches to positioning with GPS technology, in order of increasing accuracy, are as follows:

- pseudorange;
- differential carrier smoothed pseudorange; and
- differential carrier phase.

A number of variations on these approaches exist, depending on which frequencies (i.e., L1 and L2) and codes (P-code, C/A-code, and soon, Y-code, which is generated by multiplying the known P-code by the Antispoofing W-code) are used. An alternative approach involves integrating the Doppler velocity observable into the position solution. Table 3-2 summarizes the accuracies currently recovered from these approaches. Differential GPS positioning technology can provide three-dimensional positioning accuracies on the order of 2 to 20 cm under optimum conditions. Factors that affect positioning accuracies include the distance between the fixed receiver and the survey aircraft, the hardware, the propagation conditions, and the algorithms used to reduce the data.

The present limitations on the recovery of accurate positioning information for airborne applications are as follows:

- presence of Antispoofing (AS), which makes the P-code unavailable and renders the precise carrier phase solution more difficult and, subsequently, more expensive to recover;
- propagation errors due to ionospheric, tropospheric, and multipath effects;
- loss of a continuous satellite signal due to aircraft motion (cycle slips); and
- inappropriate software. (Most available software was designed for static applications and therefore is not optimized for kinematic data sets.)

The limitations of each are described below.

AS is the planned encryption of the P-code for reasons of national security (the implementation of the Y-code). P-code pseudorange measurements provide crucial information for the precise carrier phase solutions, but as Table 3-2 demonstrates, the accuracy of any solution degrades when the AS system is implemented. For example, when AS is not on, the performance level for the L2 measurements is similar to that of the L1. When AS is on, however, the signal-to-noise ratio (SNR) of the code-limited receivers typically degrades by 1,000 times and thus requires more averaging time to achieve comparable results. The implementation of AS will prevent significant advances in kinematic positioning. (A detailed discussion of the full range of effects of both AS and Selective Availability (SA) is provided in Appendix A.)

Since the fundamental measurement in precise GPS applications is the carrier phase broadcast by the receiver, any phenomenon that delays or modifies its travel time introduces errors into the final position solution. The three major sources of propagation errors are the ionosphere, the troposphere, and multipath. The use of two frequencies in the GPS system largely corrects for the dispersive delay of the ionosphere except in the polar regions where the effect of the ionosphere is greater. The atmospheric delay from space to the surface of the Earth can be as large as 2.5 m for zenith observations and 10.0 m for observations made at 15° eleva-

TABLE 3.2 Present-Day State-of-the-Art of GPS-Determined Position Accuracy* of the Airborne Platform (Moving Along a Straight Flight Path at Speeds Up To 300 Knots).

Mode	Data	SA Off AS Off	SA On AS Off	SA On or Off AS On
Positioning				
Point Positioning	L1 C/A	16 m	100 m	NA
	P-code	1-2 m	Same as C/A	Loss of P-code[a]
Differential Range	L1 C/A	3-8 m + 5 ppm[b]	Unaffected	
	L1/L2 P-code	1-2 m + 1 ppm[c]	Unaffected	
Differential Carrier Smoothed Range	L1 C/A	1-2 m + 5 ppm[b]	Unaffected	
	L1/L2 P-code	0.5-1.0 m + 3 ppm	Unaffected	Loss of P-code[a]
Carrier Phase (bias fixed)	L1 C/A	10-30 mm + 1-10 ppm	Unaffected	Unaffected
	Dual frequency	10-50 mm + 0.1 ppm	Unaffected	Loss of P-code[a]
Attitude				
Carrier Phase (10-m baseline)	L1	20" (1 s)[d]	Unaffected	Unaffected
Acceleration				
Carrier Phase	L1	200 mGal (2 s)[e]	Unaffected	Unaffected

* The range in accuracies is horizontal to vertical. A single number indicates that the horizontal and vertical accuracies are similar. Accuracies were compiled from input at 1993 Workshop on Airborne Geophysics.

[a] Loss of P-code can be overcome with a code-limited tracking scheme at the expense of loss of signal-to-noise. In dynamic applications, this loss may greatly reduce the utility of code-limited techniques. In this case, the results would revert to L1 C/A quality.

[b] For L1 only observations, the 5 ppm is taken as representative of ionospheric propagation effects. In some cases, the ionospheric contribution can be <1 ppm and in others (particularly polar and equatorial regions) >10 ppm.

[c] Limited by real-time orbit uncertainty.

[d] Assumes 1-mm differential position determination on a 10-m baseline. In severe multipath environments, results may be significantly worse.

[e] Assumes 1-mm position determinations over intervals of 2 s. Results will be affected by multipath environment.

tions. With sufficient numbers of satellites and ranges of elevation angles being observed, these delays can be estimated to an accuracy of about 10 mm. The effects of atmospheric delays are larger for higher-flying aircraft and for greater distances from the fixed ground receivers. Multipath is the phenomenon whereby the signal recorded by the receiver did not travel directly to the antennas but reflected off some nearby object such as the aircraft tail or fuselage. This is a significant, but largely undocumented source of error in the airborne kinematic positioning solution.

The loss of signal from a satellite for even a fraction of a second introduces significant errors into a precise carrier phase solution. These data outages are termed cycle slips. Recent receiver technology has greatly improved the ability of receivers to continuously track broadcast signals at the high rates required by the measurement systems. During aircraft turning maneuvers, however, the tail and the wings frequently mask the antenna's view of the satellite, resulting in numerous cycle slips.

The integration of carrier phase positioning techniques with traditional navigation systems may minimize the effect of cycle slips. The integration of GPS with INS creates the optimum navigation and positioning system, with GPS providing the primary long-term positioning system and INS providing the short-term navigation system in an entirely complementary fashion. GPS offers long-term stability and accuracy, while INS possesses short-term sensitivity and precision. Industry is actively pursuing this integration, which is hindered by the relatively high cost of INS needed for cm-level precision.

A last difficulty in the kinematic positioning of aircraft is the state of the reduction software. Typical reduction software was designed to locate isolated points on the Earth's surface, not a rapidly moving aircraft. For short-distance differential positioning (such as aircraft landings), real-time data processing can be used. The most accurate geophysical data processing, which must incorporate the large separation of aircraft and base stations, requires post-processing using precise satellite ephemeris. The data volumes for a 4-hour flight with data collected at 1 Hz at a base station and aboard an aircraft greatly exceed the capacity of these programs. The existing software is also not designed to accommodate multiple base stations, which may be necessary for long-range flights. Finally, these systems cannot easily integrate the other information collected aboard research aircraft into the position solution.

Navigation

In the absence of GPS, precise navigation applications are conducted with surface-based radio navigation networks. These systems, for example, are used widely for landing aircraft (e.g., U.S. Department of Transportation/U.S. Department of Defense, 1992). This approach can provide very accurate navigation information to an airborne platform, but it has the very great disadvantage of requiring time-consuming efforts to install a local network. Real-time GPS navigation provides a viable alternative.

Real-time positioning (navigation) to an accuracy of a few meters is needed to allow repeated surveys of the same flight lines, which are required for global change and neotectonic applications. This navigation requirement is less stringent than the positioning requirements due to the physical limitations on steering an aircraft. For example, a pilot can hold an aircraft to about 6 m of the flight line with GPS information, whereas autopilots do no better than about 10 m.

The presence of induced errors in timing that result from SA, and the inability of the civilian community to utilize P-code reliably because of AS, requires that a real-time differential solution, based on pseudorange or carrier phase technology, be employed for very precise navigation. Efforts to address these issues are being driven principally by the aviation industry's interest in implementing airport approach systems based on GPS technology (e.g., Hundley et al., 1993; Rowson et al., 1994). The navigation requirements for accurate geophysical surveys and the commercial aviation industry are very similar and pose similar problems. For both applications, communication strategies must be developed for transmitting range and phase corrections in differential GPS positioning. Data transmission rates generally must be 50 to 2,000 bits/s to ensure that the required accuracy is maintained. Instead of using base stations and communications dedicated to each survey, it would be useful to integrate commercial or government-provided differential services (e.g., through U.S. Coast Guard, FAA) with airborne geophysical surveys.

Broadcasting GPS corrections has the added potential benefit of validating real-time data. Real-time data validation techniques are important even if high-quality real-time position and velocity solutions are not needed. It is essential to determine whether the data being collected are valid so that problems can be corrected before significant flight time is wasted.

Orientation

As with positioning, various levels of attitude accuracy and temporal resolution are needed to support geophysical sensors. Typical attitude accuracies are about 1 mrad, although more stringent performance is required for interferometric SAR, photogrammetry, and particularly for vector gravimetry (<0.005 mrad). It is possible to use GPS to recover attitude by interferometry of signals received at multiple antennas over very short baselines (1 to 5 m), with up to 1 mrad in accuracy. This approach was recommended in the report *Solid Earth Sciences and Society* (NRC, 1993). In general, however, the inertial sensor is more reliable, has higher performance, and is more easily adapted to vehicle installation. It is, therefore, the recommended attitude sensor for airborne geophysics platforms.

Future Directions for Airborne GPS Technology

In light of the present limitations of GPS positioning for navigation and orientation applications, the technology should be improved in several areas. Advances should focus on overcoming the following problems:

- the implementation of AS and SA, which seriously limit the positioning and navigation applications of GPS technology;
- the source and magnitude of the propagation errors attributed to ionospheric, tropospheric, and multipath effects;
- the limitations inherent in current receiver, antenna, and communications systems;
- the kinematic GPS software, which should be improved to enhance efficiency and accuracy of positioning; and
- the integration of INS and GPS technologies to meet the stringent requirements of airborne geophysics.

Issues associated with SA and AS (see Appendix A) need to be resolved, particularly where they have a detrimental effect on airborne applications. Continuing uncertainties have led manufacturers of GPS equipment to spend considerable resources to develop technologies to counter these security systems. Such effort might have been better spent on improving tracking loops and the SNR of the received signals.

The errors introduced by multipath, thermal noise, and the temporal correlation of phase and range measurements must be adequately assessed so that reliable estimates of the noise characteristics of given scenarios can be determined. For example, statistics on how ionospheric effects in the polar and equatorial regions degrade GPS accuracy should be developed so that survey planners will recognize the occurrence and severity of these effects in a particular geographic region. Tropospheric effects can also be a problem under some conditions when the highest possible accuracy is needed, and compensating techniques should be developed.

The GPS receivers that are used to position and navigate airborne platforms should be modified to address the following deficiencies:

- Specialized antennas or processing techniques must be developed to minimize the problem of multipath on aircraft.
- The SNR on L2 codeless receiver modes must be improved to allow better performance in dynamic applications.
- Alternative tracking strategies should be developed to counter AS (Antispoofing), or AS should be turned off.

In addition, postprocessing software should be improved to become more user-friendly, to facilitate quality control, and to allow near-real-time processing and quality control.

Improvements in inertial sensor technology generally center on increasing the reliability and maintenance of the instrument, and on reducing cost, size, weight, and power requirements. Increased accuracy is not the driving factor, especially since the advent of GPS. Because of the navigation and positioning improvements afforded by integrating GPS and INS, however, it is desirable to design tightly-coupled systems containing GPS receiver electronics, micromachined silicon chip accelerometers, and fiber optic gyroscopes. In less than a decade, it is anticipated that such combined units will be no larger than 100 cubic inches, have a "mean time between failures" of 20,000 hours, and cost less than $30,000 (today's cost of a high-accuracy GPS receiver).

These improvements will meet the needs of most conventional commercial and military applications, but airborne geophysics applications may require greater accuracy and/or specialized interfaces and data accessibility (e.g., higher data rates). It is likely that the more stringent requirements for airborne geophysics will impose a higher cost, but there are no technological barriers. Even in the most demanding case, that of airborne vector gravimetry, current high-precision INS provides an adequate platform orientation to demonstrate the concept.

AIRBORNE PLATFORMS

The four airborne platforms widely used in geophysical research and development can be categorized as follows:

- discovery mission aircraft;
- process-driven mission aircraft;
- developmental aircraft; and
- experimental aircraft.

Discovery aircraft have been used for more than 20 years in the study of unknown regions. Typically, fast, long-range aircraft such as Hercules C-130s or Lockheed P-3 Orions equipped with one or two measuring systems are used. The line spacing is wide to ensure that the maximum region is covered. An example of a discovery mission is the joint U.S.-British Antarctic Survey radar flights that surveyed much of West Antarctica in the 1970s. Aircraft mapped the ice surface and bedrock topography at a 50-km line spacing and provided the first images of the wide rivers of ice that drain the West Antarctic ice sheet. More recently, discovery aircraft equipped with gravity and magnetic instruments have completed surveys over the subcontinent of Greenland.

Process-oriented aircraft are designed to study specific scientific problems on a somewhat smaller scale and with a broader range of instrumentation. Typically, process-oriented aircraft, such as the de Havilland Twin Otter, fly slower and collect data in more tightly sampled grids. For example, the CASERTZ (Corridor Aerogeophysics of the South Eastern Ross Transact Zone) program, sponsored by the National Science Foundation, was designed to investigate the interaction of ice sheets with the underlying geology. The program requires accurate measurement of the ice surface, bedrock topography, and the gravity and magnetic character of the bedrock. The track spacing is narrow (5 km) and the targeted features are on the order of 10 km.

Developmental aircraft are those used to prove a technique, demonstrate a technology, or refine a measurement; they are not targeted at furthering our understanding of the Earth's systems. Examples of this approach include using a NASA DC-8 to test SAR technology over known terrain, using small aircraft over Lake Ontario to merge GPS and INS technologies, or flying a Navy P-3 Orion over a well-characterized area to demonstrate the robustness of gravity measurements.

Experimental aircraft are platforms that are still undergoing development, such as airships (blimps) and drones. Airships may be well suited to high-resolution studies due to their slow speed and low altitude flight capabilities, but they have limited weather operating ranges and require a large ground-based crew. Unmanned aerial vehicles, or drones, are used for surveillance, reconnaissance, detection and monitoring of nuclear radiation, hazard avoidance, and data collection in flights lasting 1 to 6 hours, depending on the speed of the aircraft (Ferer and St. Pierre, 1994). Like airships, drones are susceptible to adverse weather conditions, such as turbulence, lightning, and icing, and they require somewhat elaborate ground facilities to catapult the vehicle into the air, monitor its progress, activate the parachute and shock absorber for landing, and recover the aircraft. Because drones were designed for surveillance, they have a relatively small payload capacity. Before their use in geophysical applications can become widespread, light-weight sensor packages and simplified ground control systems will need to be developed.

Each of these platforms is important in fostering a strong technological base and a large scientific user community. To apply airborne techniques to scientific problems, the research community must have access to airborne platforms. There are three primary methods for accessing any of these airborne research platforms:

1. A commercial or government contractor provides the full suite of services to the scientist, agency, or corporation.

2. A government-sponsored aircraft with government-developed equipment is provided for collaborative research.

3. A commercial or government contractor provides the aircraft, and individual research groups install the sensing and positioning equipment.

Access to discovery-oriented and process-oriented platforms is more restricted. In some cases, a contractor may be able to provide the necessary services for a discovery-oriented mission. For example, airborne magnetic services are provided commercially to geophysical researchers. However, contractor services typically focus on collecting one or two principal data types, whereas process-oriented missions have multi-instrument, interdisciplinary requirements.

Access to airborne platforms for technology development is also somewhat restricted. This use typically requires access to an aircraft for significant periods of time and is most easily accomplished within government laboratories or in joint programs between researchers and commercial or government contractors. New approaches must be considered to broaden the access of the wider user community to all of these platforms.

4

A STRATEGY FOR THE FUTURE

The previous chapters focused on the scientific directions in airborne geophysics that are possible with precise positioning and on the technological developments required to pursue them. To fulfill these research objectives, certain requirements must also be met. There needs to be

- improved access to airborne technology;
- development of flexible operational measurement systems;
- better interagency and interdisciplinary coordination; and
- reevaluation of GPS security systems.

As a result of deliberations based on the input it received, the Committee on Geodesy offers the following recommendations for meeting these requirements:

Recommendation 1. The new capabilities in precise positioning and accurate navigation should be made more accessible to the user community. An initiative to increase access should include both the establishment of an airborne earth science facility and a coordinated effort at educating its potential users.
A facility dedicated to airborne earth sciences would consist of instrumented aircraft that are made available to researchers for fixed periods of time. Such a facility must provide a broad range of services to meet the needs of the basic and applied research communities. In particular, it must promote the use of discovery, process-oriented, and development platforms. These platforms would give researchers the

capability of conducting broad regional surveys with one or two sensors, or high-resolution surveys with a broad spectrum of sensors. Technical development platforms allow new technologies to be evaluated and modified in order to develop accurate operational measurements. Access to each of these systems would provide fertile ground for scientific and technological advances.

Management of the facility, evaluation of proposals for the use of its aircraft, and efforts to educate the potential user community about its capabilities could be coordinated through a variety of management structures, including the following:

- coordinating interagency working group;
- university consortium funded by the government;
- university/government/private-sector partnership; and
- central (nonprofit) coordinating committee.

Models for each of these approaches exist throughout the government and academic community. For example, an interagency working group of nine federal agencies is pooling data, technology, and resources to create a data and information system for the U.S. Global Change Research Program. In the seismology community, Incorporated Research Institutions for Seismology (IRIS), a university consortium funded by government agencies, is responsible for distributing research funds and coordinating the installation of seismograph stations. DOE's Domestic Energy Initiative is an example of a university/government/private-sector partnership. Finally, a nonprofit coordinating committee, the Joint Oceanographic Community, coordinates the operations of the drilling ship the *Joides Resolution* for the National Science Foundation. Each of these cooperative ventures was driven by the need to access technology that is too complex, expensive, or logistically difficult to be managed by an individual investigator or agency. Airborne geophysical technology presents similar difficulties and will require the establishment of a central facility to ensure its broader use.

Existing capabilities at NASA, NSF, USGS, or the Navy could form the basis for such a government-operated facility and any number of airborne geophysical companies could provide aircraft from the private sector. As with ocean surveys, no single method of cooperation is likely to meet all scientific needs. Consequently, some missions could be flown privately and others could be flown under contract by federal agencies.

To ensure maximum use of these platforms, it is necessary to educate the potential and established user communities. Educational efforts should begin with the encouragement of collaborative projects between scientists and the federal and commercial developers of airborne geophysical technology and precise positioning. In addition, conferences that attempt to merge the small airborne geophysical community with the broader science community should be convened regularly. An appropriate forum might be a three-day Chapman Conference, sponsored by the American Geophysical Union, which would provide ample time for informal interaction and initiation of collaborative research.

Recommendation 2. Airborne geophysical measurements should be coordinated across disciplines, programs, and funding agencies to promote interdisciplinary research and to optimize use of the aircraft.
Programs in airborne geophysics are currently concentrated in a few agencies, corporations, and university departments. This segregation of effort limits both the effective use of expensive aircraft and the development of innovative research. For example, integration of gravity and magnetic measurements with major photogrammetric missions could produce significantly more data at a relatively small incremental cost. Such coordinated missions would be beneficial to both the funding agencies, which save money, and to the broader scientific community, which gains data and the opportunity to do interdisciplinary research.

Recommendation 3. To ensure uniform coverage that is sufficiently accurate to resolve both long- and short-wavelength geologic features, technological developments should aim at integrating GPS with a broad spectrum of well-calibrated measurement systems.
Uniform, accurate measurements are needed for both long-wavelength regional studies and short-wavelength process-oriented studies. Future technological developments must address both needs to ensure that airborne geophysical methods become routine tools for the scientific, resource, and environmental industries. Currently, only a limited number of accurate operational systems are routinely available. These include high-resolution airborne magnetics and photogrammetry, which are commercially available, and airborne gravity, which is available through

government, industry, and academia. Topographic mapping systems are still largely in the developmental stage and are not available for routine scientific operations.

Other technological advances should focus on the design and implementation of experiments that require multiple measurement systems on a single platform. These applications require well-calibrated, integrated operational systems. The design process should look beyond the proof-of-concept and developmental stages to the practical considerations of an aircraft survey. Integrated operational systems will place specific restrictions on equipment size, power requirements, and survey design.

Recommendation 4. In light of the serious impact on airborne geophysics, particularly for emerging industrial applications, the continuous operation of the Antispoofing system should be carefully evaluated.

The applications and technological advances discussed in previous chapters depend on access to GPS signals that make it possible to locate the aircraft's antenna to better than 1 m. When AS is on, these applications can be accomplished only if other parts of the positioning system are improved. Such improvements include replacing all but one manufacturer's existing GPS receivers with models that use a different encoded signal for positioning; installing additional ground tracking systems so that aircraft are never more than tens of kilometers away from a ground station; or coupling the GPS receiver to the inertial guidance system in the aircraft to account for the dynamics of the aircraft. Even when AS is off, very high resolution applications will require improved navigation and postmission processing. The implementation of the AS system, however, would seriously hamper improvements in these technologies. The continued use of AS must be evaluated in light of these very real civilian consequences.

APPENDIX A

EFFECTS OF SELECTIVE AVAILABILITY (SA) AND ANTISPOOFING (AS)

Selective Availability (SA) denies precise positioning by corruption of the GPS signal structure. It is composed of two components: (1) corruption of broadcast navigation message; and (2) rapid "dithering," or oscillations of the frequency standards in the satellites. The first component of SA is of little consequence to scientific users because in precise millimeter applications, the orbits of the GPS satellites are computed from carrier measurements much more accurately than even the precise ephemeris available from the U.S. Department of Defense. Also, the dynamics of the GPS satellites are understood well enough that these orbits can be integrated forward in time by several days with accuracies better than the (uncorrupted) broadcast ephemeris, except when there are thruster firings on the GPS satellites.

The second SA component also currently has little effect when differential (either real-time or postprocessing) techniques are used (Feigl et al., 1991). Figure A.1 provides an example of the dithering of the GPS frequency standards obtained from analysis of the data collected with the global GPS tracking network. In this example, SA is turned off on the satellite at approximately 18:00 Universal Time Coordinated (UTC). The results of the dithering have been converted to meters of range error in this figure. Before SA is turned off, the root-mean-square (RMS) error in the range measurements is 22 m; when SA is turned off, the range error RMS drops to about 0.25 m. (These results are obtained from the analysis of carrier phase measurements, and therefore nearly all of the RMS is due to dithering and natural drifts in the satellite and ground station clocks.) While the effects of SA are almost totally eliminated in differential positioning, the rapid fluctuations in the satellite clocks complicate the process of removing slips in the number of carrier phase cycles accumulated by the receiver. In addition, SA makes it nearly impossible to use averaged phase measurements (normal points) with a much lower sampling

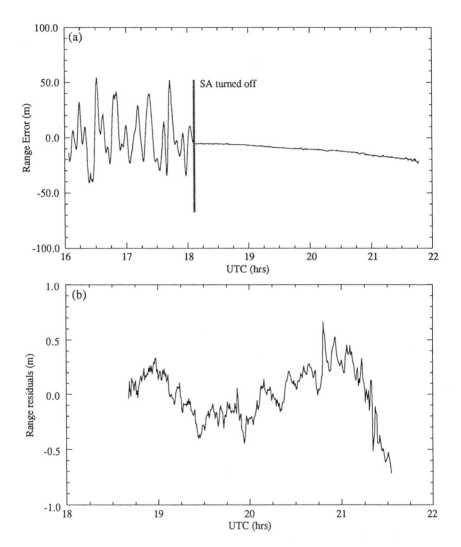

FIGURE A.1 Example of the effect of Selective Availability. In this example, SA was turned off at about 18:00 UTC, and the impact of SA can be clearly seen. In (a), the range error to GPS satellite number 1 is shown. The RMS error in the range errors when SA is on is 22 m, consistent with denying point positioning accuracies better than 100 m. In (b), an expanded view of the time interval when SA is not on is shown after a quadratic polynomial is removed. These results were obtained using carrier phase measurements. The RMS range error for this interval is 0.25 m, and indicates that 1-m accuracy point positioning would be possible with existing satellites and receivers if SA was not on.

rate.   In real-time navigation problems, SA limits the accuracy of navigation, although this is probably not a major problem for aircraft since the position of the aircraft can generally only be controlled to within about 100 m, the positional accuracy imposed by SA.

Antispoofing (AS) is meant to stop false signals from corrupting military receivers, but, as a consequence of the system used, AS denies access to the P-code.   AS is implemented by modulating the P-code with an additional code (called the W-code), and because only the P-code (with the W-code modulation) is superimposed on the L2 frequency, some type of codeless tracking of L2 is required when AS is turned on.   Figure A.2 shows the effect on two different receivers when AS is turned on at approximately 17:00 UTC.   The degradation of the performance of the receiver in Figure A.2(a) is immediately obvious and severely limits the effectiveness of the automatic data processing systems.   In Figure A.2(b), the degradation of the performance of the receiver is not as obvious, although careful analysis of these data indicates that the effectiveness of automatic data processing systems is also compromised when AS is turned on.   (It is also interesting to note the general stability of the receiver shown in Figure A.2(b) is not as good as that for the receiver in Figure A.2(a) when AS is off, possibly indicating the manufacturer of receiver (b) has put most of the efforts into handling AS rather than developing a stable receiver for the periods when AS is off.)   The prime consequence of AS is the loss of accuracy in both range and phase measurements.   For aircraft applications, this loss of accuracy (about a thousandfold for L2 tracking) is particularly severe because of the dynamics of the aircraft and the usually high multipath environment on the aircraft.   In particular, results of the quality shown in Figure A.2(b) can only be obtained by coherently averaging the L2 signal for about 1 s.   This is an acceptable compromise for a static receiver, but in an aircraft it often leads to loss of lock on the L2 signal.

When a GPS receiver loses lock on a satellite, an arbitrary number of cycles of discontinuity in the carrier phase are introduced to the data.   The postprocessing removal of cycle slips is complicated by the poorer-quality pseudorange data available under AS conditions.   Further complications arise from the time it takes the receiver to reacquire lock on the satellite.   Under AS conditions, reacquiring lock can take 5 to 10 seconds (of order 1 km of flight path).   During this time, no L2-range data are available, resulting in gaps in the flight path when dual-frequency positions are

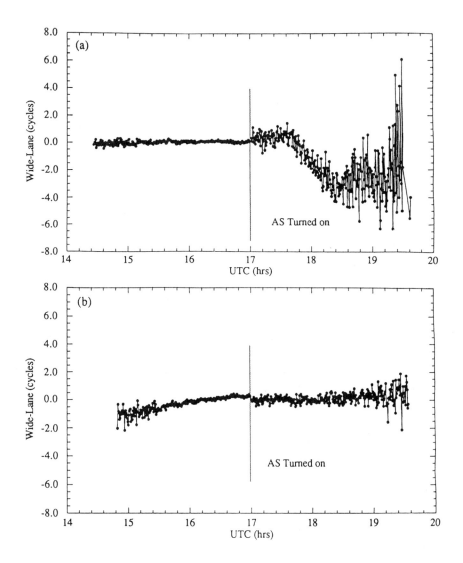

required (during conditions of significant ionospheric delay and when the aircraft is far from its ground tracking station). Improvements in current processing algorithms can improve the robustness of the analysis for interpolating across these times if the occurrences are infrequent enough. When losses of lock are too frequent under AS conditions, the utility of dual-frequency GPS will be totally lost and accurate aircraft positions will not be available.

In principle, the effects of AS on signal frequency (L1) observations should be minor, although the effect can depend on how the P-code is treated on the L1 frequency (e.g., some receivers remove the P-code from the L1 signal before cross-correlating with the C/A code; such techniques will not work correctly when AS is active). However, because the P-code is written with lower power than the C/A code and the codes decorrelate rapidly, the presence of the P-code on the L1 signal can be largely ignored in the extraction of the C/A code. When the P-code is ignored, AS has no effect. Some dual-frequency GPS receivers couple the L1 and L2 tracking, and these receivers can experience loss of lock on the L1 signal when there is a loss of lock on the L2 signal. Improvements to the tracking loops in these receivers could improve this characteristic.

FIGURE A.2 Example of the effects of Antispoofing. In this example, AS is turned on at 17:00 UTC and the effects on the receiver shown in (a) are dramatic. The quantity plotted, the widelane, is a linear combination of L1 and L2 phase and range measurements that is independent of both geometry variations and the ionospheric delay. It is a direct measure of the difference in the number of cycles between the L1 and L2 phase measurements, and it is essential for rapid-static applications of GPS. Before AS is turned on, the RMS scatter of widelane, which needs to be small compared to one cycle if the widelane is to be useful, is 0.06 cycles (11 mm). For the first hour after AS is turned on, the RMS scatter jumps to 0.7 cycles (133 mm) and degrades even more as the satellite moves to low-elevation angles and the signal-to-noise of the GPS measurements decreases.

In (b), the results obtained with another (later-generation) receiver making measurements at the same time are shown. Here the change in RMS when AS is turned on is not so large; 0.15 cycles (29 mm) before, and 0.20 cycles (40 mm) afterward. However, the similarity before and after is deceptive because when AS is turned on, the averaging times of the measurements need to be increased a thousandfold, which is effective for the static case shown but could have severe effects in dynamic applications of GPS.

APPENDIX B

CONCLUSIONS FROM THE WORKSHOP ON AIRBORNE GEOPHYSICS

The Workshop on Airborne Geophysics was held to assess the promise of airborne geophysical techniques and precise positioning in supporting future scientific research in regional geodesy, from surveying applications to the measurement of global warming indicators, such as glacier and ice sheet mass balance. The workshop addressed the following themes:

- GPS positioning;
- integrating GPS and Inertial Navigation Systems (INS);
- recovery of topography from aircraft;
- airborne gravity field measurements; and
- scientific objectives.

Following are the major conclusions of the workshop:

Workshop Conclusion 1: Current airborne geophysics technology is sufficiently mature to address basic research problems and commercial applications in both the resource and environmental fields. Increased visibility and access to the present systems would encourage wider application of airborne geophysics integrated with precise positioning.

Workshop Conclusion 2: Precise aircraft positioning capability with GPS should motivate development of techniques for swath mapping topography with an accuracy between 10 and 100 cm and the capability to detect changes in topography in the range of 1 to 10 cm. This would allow advances in profiling ice sheets, resolving the horizontal and vertical response of the Earth's surface following a major seismic event, detecting magma chamber swelling, and verifying nuclear test ban treaties.

Workshop Conclusion 3: Airborne gravity technology is currently capable of detecting anomalies with wavelengths of several kilometers and accuracies of several milligal. Increasing the accuracy and resolution of airborne gravity measurements to the sub-mGal and less than 1-km range should be pursued by enhancing precise positioning capabilities, improving sensors, and developing new technologies such as gravity gradiometers. These high-resolution applications are of particular importance for mineral and petroleum resource evaluation.

Workshop Conclusion 4: Research and development in GPS and INS should more directly target high-accuracy aircraft applications. Much of the available hardware and software was designed specifically for static applications; airborne applications will require more attention on multipath problems, tracking loops, and reduction software.

Workshop Conclusion 5: The implementation of Antispoofing policies, which increase the difficulty in tracking the GPS signal, seriously encumbers future applications of airborne geophysics.

APPENDIX C

WORKSHOP ON AIRBORNE GEOPHYSICS:
THE AGENDA

July 12-14, 1993

National Research Council
Green Building, Room 130
2001 Wisconsin Avenue, NW
Washington, DC

July 12   *GPS Positioning/Navigation and GPS/INS Integration*

Session 1

8:00   Registration/Continental breakfast

8:30   Welcome   Bell

8:40   **Scientific Framework for Airborne Geophysics**
       (chair: Minster)   McNutt

9:30   **Federal Agency Contribution:**   Agency liaisons
       Programmatic overview
       Baltuck (NASA)
       Battis (AF)
       Borg/Peacock (NSF)
       Hanna (USGS)
       Smith (DMA)

11:00 Coffee break

Session 2

11:15 **Precise Positioning and Navigation with GPS**
      (chair: Sailor)   Herring

12:00 Two-minute commercials for posters   Panelists

        Lachapelle: *GPS carrier phase receiver noise and multipath estimation*
        Mader: *Recent advances in kinematic GPS processing*
        Yunck: *A fast technique for computing precise aircraft acceleration from GPS phase*
        Cannon: *'On-the-fly' ambiguity resolution: Single and dual frequency results*
        Goad: na
        Childers: *GPS software comparison and precise surface altimetry*

12:15 Lunch

1:00 Posters

1:45 Discussion and questions

Session 3

2:15   **Integration of GPS and INS**   Schwarz
         (chair: Jekeli)

3:00   Refreshments

3:15   Two-minute commercials for posters   Panelists

Brown:	*Integrated GPS/inertial mapping system for real-time and post-test precision positioning*
Greenspan:	*GPS/inertial integration overview*
Hein:	*High-precision aircraft navigation using DGPS/INS integration*
Huddle:	*The evolution of medium accuracy inertial system technology*
Webb:	*Aspects of the application of GPS to topographic mapping with airborne interferometric Synthetic Aperture Radar*

3:45   Posters

4:30   Discussion and questions

5:00   Adjourn

Evening Session

7:30   Panels 2 (Room 118) and 3 (Room 114) meet; prepare summary of sessions and draft of recommendations

July 13   **Techniques and Applications**

Session 4

8:30   Continental breakfast

9:00   **Airborne Swath Imaging**
      (chair: Sandwell)
      Scientific applications of airborne SAR interferometry   Dixon
      Kinematic GPS photogrammetry   Lucas

10:00 Coffee break

10:15 Two-minute commercials for posters   Panelists

        Hothem: *Airborne GPS control for the National Aerial Photography Program*
        Malhotra: *Simulation study for integrating GPS-control and infrared photography in aerotriangulation for shoreline mapping*
        Bock: *Synthetic aperture radar in support of crustal deformation monitoring in Sumatra, Indonesia*
        Bindschadler: *Photoclinometry of ice sheets using airborne radar altimetry as control*
        Raymond: *Integrating GPS and SAR for efficient airborne topographic mapping*

11:45 Discussion and questions

12:15 Lunch

Session 5

1:30	**Airborne Geophysical Profiling**	Blankenship
	(chair: Bell)	

2:00	**GPS and Airborne Geophysics**	Brozena

2:30	Two-minute commercials for posters	Panelists

Cordell: *Expectations for airborne gravity/gravity-*
           gradiometry in support of geologic mapping
Harding: *Airborne laser altimetry techniques for*
           geoscience studies
Ware: *Atmospheric noise in airborne gravimetry*
Gumert: *High resolution aerogravity measurements*
Ridgway: *ERS1/Airborne radar comparison over*
           southern Greenland

3:00   Posters/Refreshments

4:00   Discussion and questions

4:30   Panels meet; prepare summary of sessions

5:00   Adjourn

Evening Session

7:30   *Panels 4 (Room 118) and 5 (Room 114) meet; prepare summary*
       of sessions and draft of recommendations

July 14   **Impact of Airborne Techniques on Geophysics/Geodesy**

Session 6

7:30   Panel chairs meet to write preliminary recommendations
        (Room 116)

8:15   Continental breakfast

8:45   **Impact of Airborne Techniques on Geophysics/Geodesy**
        (chair: Herring)
        Future directions of airborne techniques   Dozier

9:45   Coffee break

10:00   Two-minute commercials for posters   Panelists

Schutz:	*GPS application to satellite altimetry: Current and future*
Bilham:	*A proposed new Freefall Inertial Gravity Gradiometer (FIGG)*
Melbourne:	*The GPS flight experiment on TOPEX/ Poseidon*
Pridmore:	*na*
Rundle:	*Airborne geophysical observations of Long Valley Caldera*
Paik:	*Application of a superconducting gravity gradiometer to airborne survey*

10:30   Posters/Coffee

11:30   Discussion and questions

12:15   Lunch

Session 7

1:00   **Plenary Session**
       Recommendations and scientific priorities
          (chair: Minster)   Panel chairs

2:30   Workshop adjourns

3:00   **Executive Session**: Steering Committee,
       Panel chairs, & COG members

5:00   Adjourn

Appendix D

Miriam Baltuck
National Aeronautics and
 Space Administration
Solid Earth Science Branch,
SEP-05, Room 5080
Earth Science and
 Applications Division,
300 E Street, S.W.
Washington, DC 20546
T: (202) 358-0273
F: (202) 358-3098
mbaltuck@nasamail.nasa.gov

James Battis
Terrestrial Science Division
Air Force Geophysical Lab
Hanscom Air Force Base
Bedford, MA 01731
T: (617) 377-3486
F: (617) 377-2707
battis@pl9000.af.mil

Robin E. Bell
Lamont-Doherty Earth
 Observatory
Columbia University, RT 9W
Palisades, NY 10964
T: (914) 365-8827
F: (914) 365-0718
robinb@lamont.leo.columbia.edu

Roger Bilham
Department of Geological Sciences
Campus Box 250
University of Colorado
Boulder, CO 80309
T: (303) 492-6189
bilham_r@cubldr.colorado.edu

Robert Bindschadler
NASA/Goddard Space Flight
 Center
Code 971
Greenbelt, MD 20771
T: (301) 286-7611
F: (301) 286-2717
bob@laural.gsfc.nasa.gov

Donald Blankenship
Institute for Geophysics
University of Texas, Austin
Austin, TX 78713
T: (512) 471-0489
F: (512) 471-8844

Yehuda Bock
Scripps Institution of
 Oceanography
IGPP A025
University of California, San Diego
La Jolla, CA 92093
T: (619) 534-5292
F: (619) 534-8090
bock@bull.ucsd.edu

Scott Borg
Program Manager for Polar
 Earth Science
Division of Polar Programs
National Science Foundation
1800 G Street, NW, Room 620
Washington, DC 20550
T: (202) 357-7766
F: (202) 357-9422

Alison Brown
NAVSYS Corporation
14960 Woodcarver Road
Colorado Springs, CO 80921
T: (719) 481-4877
F: (719) 481-4908

John Brozena
Naval Research Laboratory
316 BG 1
4555 Overlook Ave., SW
Washington, DC 20375
T: (202) 767-2024
F: (202) 767-0167
john@hp8c.nrl.navy.mil

M. Elizabeth Cannon
Dept of Geomatics Engineering
University of Calgary
2500 University Drive, NW
Calgary, Alberta T2N 1N4
Canada
T: (403) 220-3593
F: (403) 284-1980

Charles Carrigan
Lawrence Livermore National Lab
Earth Science Department, L201
Livermore, CA 94550
T: (510) 422-3941
carrigan@s94.es.llnl.gov

Vicki A. Childers
Lamont-Doherty Geological
 Observatory
Columbia University, RT 9W
Palisades, NY 10964
T: (914) 359-2900
F: (914) 365-0718
vicki@lamont.leo.columbia.edu

Lin Cordell
U.S. Geological Survey
927 National Center
Reston, VA 22092
T: (703) 648-6379
F: (703) 648-4828

Kari Craun
U.S. Geological Survey
National Mapping Division
510 National Center
Reston, VA 22092
T: (703) 648-4658
F: (703) 648-5585
kcraun@usgs.gov

Stanley Dickinson
Terrestrial Sciences
Air Force Office of Scientific
 Research/NL
Bolling Air Force Base
Washington, DC 20332-0001
T: (202) 767-5021
F: (202) 404-7475
F: (703) 695-4580 (Pentagon)

Timothy H. Dixon
University of Miami
Rosenstiel School of Marine and
 Atmospheric Sciences
4600 Rickenbacker Causeway
Miami FL 33149-1098
T: (305) 361-4660

Jeffrey Dozier
Center for Remote Sensing and
 Environmental Optics
Univ of California, Santa Barbara
Santa Barbara, CA 93106
T: (805) 893-2309
F: (805) 893-2578
dozier@eos.ucsb.edu

Jeff Genrich
Scripps Institution of Oceanography
University of California, San Diego
La Jolla, CA 92093
T: (619) 534-2661

Clyde C. Goad
Ohio State University
1958 Neil Avenue
Columbus, OH 43210
T: (614) 292-7118
F: (614) 292-2957
cgoad@magnus.acs.ohio-state.edu

Richard L. Greenspan
CS Draper Lab, Inc., MS-70
555 Technology Sq.
Cambridge, MA 02139
T: (617) 258-4041
F: (617) 258-4444

William Gumert
Carson Geoscience Co.
32H Blooming Glen Road
Perkasie, PA 18944
T: (215) 249-3535
F: (215) 249-1352

William F. Hanna
U.S. Geological Survey
927 National Center
Reston, VA 22092
T: (703) 648-6362
F: (703) 648-6684
whanna@bgrdg1.er.usgs.gov

David Harding
NASA/Goddard Space Flight Center
Code 921
Greenbelt, MD 20771
T: (301) 286-4849
F: (301) 286-1616
harding@denali.nasa.gsfc.gov

Gunter W. Hein
University FAF Munich
Institute of Astronomical
 & Physical Geodesy
Werner-Heisenberg-Weg 39
D-85577 Neubiberg, Germany
T: 49-89-6004-3425
F: 49-89-6004-3019
ac1bhein@rz.unibw-muenchen.de

Jim Heirtzler
NASA/Goddard Space Flight Center
Greenbelt, MD 20771
T: (301) 286-8364
F: (301) 286-9200

Thomas Herring
Dept of Earth, Atmospheric,
 and Planetary Sciences
Massachusetts Institute of
 Technology
Cambridge, MA 02139
T: (617) 253-5941
F: (617) 253-1699
tah@prey.mit.edu

Larry Hothem
U.S. Geological Survey
510 National Center
Reston, VA 22092
T: (703) 648-4663
F: (703) 648-5585
hothem@usgs.gov

James Huddle
Litton Guidance & Control
 Systems
MS-67
5500 Canoga Avenue
Woodland Hills, CA 91365
T: (818) 715-3264

Chris Jekeli
Dept of Geodetic Science
 and Surveying
Ohio State University
1958 Neil Avenue
Columbus, OH 43210
T: (614) 292-7117
F: (614) 292-2957
cjekeli@magnus.acs.ohio-state.edu

Peter Krueger
Headquarters TI, MS A14
Defense Mapping Agency
8613 Lee Highway
Fairfax, VA 22031-2139
T: (703) 285-9236
F: (703) 285-9396

John La Brecque
NASA Headquarters
SEP
300 E Street, SW
Washington, DC 20546
T: (202) 358-0273
F: (202) 358-3098
jlabrecq@se.hq.nasa.gov

Gerald Lachapelle
Department of Geomatics Engineering
University of Calgary
2500 University Drive, NW
Calgary, Alberta T2N IN4
Canada
T: (403) 220-7104
F: (403) 284-1980

Robert Lees
Horizons Technology, Inc.
1725 Duke Street, Suite 610
Alexandria, VA 22314
T: (703) 684-5850

Craig S. Lingle
Geophysical Institute
University of Alaska
Fairbanks, AK 99775-0800
T: (907) 474-7679
F: (907) 474-7290
clingle@iias.images.alaska.edu

Anne Linn
National Research Council
2001 Wisconsin Ave., NW
Washington, DC 20007
T: (202) 334-2744
F: (202) 334-1377
alinn@nas.edu

James Lucas
National Oceanic and Atmospheric
 Administration
SSMC3, Station 4734
1315 East West Highway
Silver Spring, MD 20910
T: (301) 713-2650
F: (301) 713-4581

Ian MacGregor
National Science Foundation
1800 G Street, NW
Washington, DC 20550
T: (202) 357-9591
F: (202) 357-0364

Gerald L. Mader
National Oceanic and Atmospheric
 Administration
National Geodetic Survey
N/CG114 NOS
Rockville, MD 20852
T: (301) 713-2854
F: (301) 468-5714

Roop Malhotra
National Oceanic and Atmospheric
 Administration
Rockville, MD 20852
T: (301) 713-2648

Maria Marsella
U.S. Geological Survey
510 National Center
Reston, VA 22092
T: (703) 648-4595

Marcia K. McNutt
Department of Earth, Atmospheric,
 and Planetary Sciences
Massachusetts Institute of
 Technology, 546-826
Cambridge, MA 02139
T: (617) 253-7304
F: (617) 253-6208
mcnutt@athena.mit.edu

William G. Melbourne
Jet Propulsion Laboratory
MS 238-540
4800 Oak Grove Drive
Pasadena, CA 91109
T: (818) 354-5071
F: (818) 393-6686
[wmelbourne/jpl]telemail

J. Bernard Minster
Scripps Institution of Oceanograph
IGPP A025
University of California, San Dieg
La Jolla, CA 92093
T: (619) 534-5650
F: (619) 534-2902 or 5332
jbminster@ucsd.edu

Martin Moody
Department of Physics
University of Maryland
College Park, MD 20742
T: (301) 405-6095
F: (301) 405-6087

Ho-Jung Paik
Department of Physics
University of Maryland
College Park, MD 20742
T: (301) 405-6086
F: (301) 405-6087

Dennis Peacock
National Science Foundation
1800 G Street, NW
Washington, DC 20550
T: (202) 357-7766
F: (202) 357-7894
dpeacock@nsf.gov

Don Pridmore
World Geoscience
17 Emerald Terrace
West Perth
Western Australia 6005
T: (713) 647-9025
T: (09) 383-7833
F: (09) 383-7166

Carol A. Raymond
Jet Propulsion Laboratory
MS 183-501
4800 Oak Grove Drive
Pasadena, CA 91109
T: (818) 354-8690
F: (818) 393-5059
car@orion.jpl.nasa.gov

Jeff Ridgway
Scripps Institution of Oceanography
University of California, San Diego
La Jolla, CA 92093
T: (619) 534-2064

John B. Rundle
Cooperative Institute for Research in
  Environmental Sciences
University of Colorado
Campus Box 216
Boulder, CO 80309-0216
T: (303) 492-1143
F: (303) 492-1149

Richard Sailor
TASC, Inc.
55 Walkers Brook Drive
Reading, MA 01867
T: (617) 942-2000, ext. 2226
F: (617) 942-7100
rvsailor@tasc.com

David Sandwell
Scripps Institution of Oceanography
University of California, San Diego
La Jolla, CA 92093
T: (619) 534-7109
F: (619) 534-0784
sandwell@radar.ucsd.edu

Bob E. Schutz
Department of Aerospace
  Engineering
WRW 402
University of Texas at Austin
Austin, TX 78712
T: (512) 471-4267
F: (512) 471-3570
schutz@utcsr.ae.utexas.edu

Klaus-Peter Schwarz
Dept of Geomatics Engineering
University of Calgary
2500 University Drive, N.W.
Calgary, Alberta T2N 1N4
Canada
T: (403) 220-7377
F: (403) 284-1980

Randall W. Smith
Defense Mapping Agency
8613 Lee Highway
Fairfax, VA 22031
T: (703) 285-9236
F: (703) 285-9396

Randolph Ware
UNAVCO
P.O. Box 3000
Boulder, CO 80309
T: (303) 497-8005
F: (303) 497-8028
ware@unavco.ucar.edu

Frank Webb
Jet Propulsion Laboratory
MS 238-625
4800 Oak Grove Drive
Pasadena, CA 91109
T: (818) 354-4670
fhw@cobra.jpl.nasa.gov

William Young
Riverside Flood Control and
  Water Conservation District
1995 Market Street
P.O. Box 1033
Riverside, CA 92501
T: (909) 275-1223
F: (909) 788-9965

Thomas Yunck
Jet Propulsion Laboratory
MS 238-640
4800 Oak Grove Drive
Pasadena, CA 91109
T: (818) 354-3369
F: (818) 393-4965

ACRONYMS

AF	Air Force
AS	Antispoofing
CASERTZ	Corridor Aerogeophysics of the South Eastern Ross Transect Zone
COG	Committee on Geodesy
DEM	Digital Elevation Model
DMA	Defense Mapping Agency
GPS	Global Positioning System
NASA	National Aeronautics and Space Administration
NSF	National Science Foundation
RMS	Root Mean Square
SA	Selective Availability
SAR	Synthetic Aperture Radar
SNR	Signal-to-Noise Ratio
USGS	U.S. Geological Survey
UTC	Universal Time Coordinated

REFERENCES

Argand, E., 1924, La tectonique de l'Asia. Proceeding of the 13th International Geological Congress, Brussels, p. 171-372.

Babu, H.V.R., N.K. Rao, and V.V. Kumar, 1991, Bedrock topography from magnetic anomalies: An aid for groundwater exploration in hard-rock terrains. Geophysics, v. 56, p. 1051-1054.

Baljinnyam, I., A. Bayasgalan, B.A. Borisov, A. Cisternas, M.G. Dem'yanovich, L. Ganbaatar, V.M. Kochetkov, R.A. Kurushin, P. Molnar, H. Philip, and Yu.Ya. Vashchilov, 1993, Ruptures of major earthquakes and active deformation in Mongolia and its surroundings. Geological Society of America Memoir 181, 62 pp.

Band, L.E., and E.F. Wood, 1988, Strategies for large-scale hydrologic simulation. Applied Mathematics and Computation, v. 27, p. 23-37.

Blankenship, D.D., R.E. Bell, V.A. Childers, and S.M. Hodge, 1992, Airborne measurement of ice-sheet elevation. EOS, Transactions of the American Geophysical Union, v. 73, p. 129.

Blankenship, D.D., R.E. Bell, S.M Hodge, J.M. Brozena, J.C. Behrendt, and C.A. Finn, 1993, Active volcanism beneath the West Antarctic ice sheet and implications for ice-sheet stability. Nature, v. 361, p. 526-529.

Bodard, J.M., J.G. Creer, and M.W. Asten, 1993, Next generation high resolution airborne gravity reconnaissance in oil field exploration. Energy Exploration and Exploitation, August Special Issue, p. 198-234.

Burke, K., and T.H. Dixon, 1988, Topographic Science Working Group Report. Washington, DC: Land Processes Branch, National Aeronautics and Space Administration, 64 pp.

Brozena, J., M. Chaloma, R. Forsberg, and G. Mader, 1992, The Greenland Aerogeophysics Project. EOS, Transactions of the American Geophysical Union, v. 73, p. 130.

Brozena, J.M., G.L. Mader, and M.F. Peters, 1989, Interferometric Global Positioning System: Three-dimensional positioning source for airborne gravimetry. Journal of Geophysical Research, v. 94, p. 12153-12162.

Cannon, M.E., 1991, Airborne GPS/INS with an application to aerotriangulation. UCSE Report 20040, Department of Geomatics Engineering, The University of Calgary, 171 pp.

Childers, V.A., R.E. Bell, D.D. Blankenship, and J.M. Brozena, 1992, Vertical aircraft positioning for gravity and precise altimetry. EOS, Transactions of the American Geophysical Union, v. 73, p. 129.

Cohen, C.E., B.S. Pervan, D.G. Lawrence, H.S. Cobb, J.D. Powell, and B.W. Parkinson, 1994, Real-time flight testing using integrity beacons for GPS Category III precision landings. Navigation, v. 41, p. 145-157.

Coolfont, P.A., 1991, Solid Earth Sciences in the 1990's. Volume 2, Panel Reports. NASA Technical Memorandum 4256, Washington, DC: Office of Space Science Applications, National Aeronautics and Space Administration, 279 pp.

Dixon, T., E.R. Ivins, and B.J. Franklin, 1989, Topographic and volcanic asymmetry around the Red Sea: Constraints on rift models. Tectonics, v. 8, p. 1193-1216.

Doll, W.E., J.E. Nyquist, J.S. Holladay, V.F. Labson, and L. Pellerin, 1993, Preliminary results of a helicopter electromagnetic and magnetic survey of the Oak Ridge Reservation, Tennessee for environmental and geologic site characterization. *In* Bell, R.S., and C.M. Lepper, eds., Proceedings of the Symposium on the Application of Geophysics to Engineering and Environmental Problems, p. 281-295.

Enge, P.K., D. Young, L. Sheynblatt, and B. Westfall, 1994, DGPS/radiobeacon field trials, comparing Type 1 and Type 9 messaging. Navigation, v. 40, p. 395-408.

Evans, D.L., T.G. Farr, H.A. Zebker, J.J. vanZyl, and P.J. Mouginis-Mark, 1992, Radar interferometry studies of the Earth's topography. EOS, Transactions of the American Geophysical Union, v. 73, p. 553, 557-558.

Feigl, K.L., R.W. King, T.A. Herring, and M. Rothacher, 1991, A scheme for reducing the effect of Selective Availability on precise geodetic measurements from the Global Positioning System. Geophysical Research Letters, v. 18, p. 1289-1292.

Ferer, K.M., and D.B. St. Pierre, 1994, European unmanned aerial vehicles - Overview. European Naval Oceanography, Office of Naval Research Europe Newsletter, August 17.

Fletcher, R.C., and B. Hallet, 1983, Unstable extension of the lithosphere; a mechanical model for Basin and Range structure. Journal of Geophysical Research, v. 88, p. 7457-7466.

Frodge, S.L., B. Remondi, and D. Lapucha, 1994, Results of real-time testing and demonstration of the U.S. Army Corps of Engineers real-time on-the fly positioning system. Proceedings of the XXth Congress of the International Federation of Surveyors, Melbourne, March 5-12, 1994, v. 6, TS 606.3.

Fu, L.L.R., 1983, Recent progress in the application of satellite altimetry to observing the mesoscale variability and general circulation of the oceans. Reviews of Geophysics, v. 21, p. 1657-1666.

Garvin, J.B., 1993, Mapping new and old worlds with laser altimetry. Photonics Spectra, v. 27, p. 89-94.

Garvin J.B., and R.S. Williams Jr., 1990, Small domes on Venus: Probable analogs of Icelandic lava shields. Geophysical Research Letters, v. 17, p. 1381-1384.

Grauch, V.J.S., D.A. Sawyer, M.R. Hudson, S.A. Minor, and J.C. Cole, 1993, New, detailed aeromagnetic data give fresh insight to mapping covered geologic units in the southwest Nevada volcanic field. EOS, Transactions of the American Geophysical Union, v. 74, p. 221.

Guignard, J.P., 1992, ERS-1 SAR Calibration Strategy. Proceedings of the First ERS-1 Symposium, Cannes France, November 4-6, 1992. European Space Agency, Paris, France, p. 151-171.

Gumert, W.R., 1992, Airborne gravity measurements. In Geyer, R.A., ed., CRC Handbook of Geophysical Exploration at Sea, 2nd edition, Hydrocarbons, Boca Raton: CRC Press, p. 125-139.

Harding, D.J., J.B. Blair, J.B. Garvin, and W.T. Lawrence, 1994, Laser altimetry waveform measurement of vegetation canopy structure. Proceedings of the International Geoscience and Remote Sensing Symposium, Pasadena, CA., v. 2, p. 1251-1253.

Heiskanen, W.A., and H. Moritz, 1967, Physical Geodesy. San Francisco: W.H. Freeman, 364 pp.

Hofmann-Wellenhof, B., H. Lichtenegger, and J. Collins, 1992, GPS Theory and Practice, New York: Springer Verlag, 326 pp.

Houser, F.N., 1970, A summary of information and ideas regarding sinks and collapse, Nevada Test Site. U.S. Geological Survey Open File Report 474-41 (NTS-216).

Hundley, W., S. Rowson, G. Courtney, V. Wullschleger, R. Velez, and P. O'Donnel, 1993, Flight evaluation of a basic C/A code differential GPS landing system for Category I precision approach. Navigation, v. 40, p. 161-178.

Jekeli, C., 1988, The Gravity Gradiometer Survey System (GGSS). EOS, Transactions of the American Geophysical Union, v. 69, p. 105, 116-117.

Jekeli, C., 1993, A review of Gravity Gradiometer Survey System data analyses. Geophysics, v. 58, p. 508-514.

Jin, Y., M.K. McNutt, and Y. Zhu, 1994, Evidence from gravity and topography data for folding of Tibet. Nature, v. 371, p. 669-674.

Labson, V.F., A.T. Mazzella, and G.J. Schneider, 1993, Airborne geophysical survey for environmental site characterization. EOS, Transactions of the American Geophysical Union, v. 74, p. 220.

Madsen, S.N., H.A. Zebker, and J. Martin, 1993, Topographic mapping using radar interferometry: processing techniques. Transactions on Geoscience and Remote Sensing, v. 31, p. 246-255.

Massonnet, D., M. Rossi, C. Carmona, F. Adragna, G. Peltzer, K. Feigl, and T. Rabaute, 1993, The displacement field of the Landers earthquake mapped by radar interferometry. Nature, v. 364, p. 138-142.

Meier, M.F., 1984, Contribution of small glaciers to global sea level. Science, v. 226, p. 1418-1421.

Mercer, J.H., 1978, West Antarctic ice sheet and CO_2 greenhouse effect; a threat of disaster. Nature, v. 271, p. 321-325.

Milbert, D.G., 1991, GEOID90: A high-resolution geoid for the United States. EOS, Transactions of the American Geophysical Union, v. 72, p. 545-554.

Molnar, P., and P. Tapponnier, 1975, Cenozoic tectonics of Asia; effects of a continental collision. Science, v. 189, p. 419-426.

National Aeronautics and Space Administration (NASA), 1987, Geophysical and geodetic requirements for global gravity field measurements, 1987-2000. Report of a Gravity Workshop, Colorado Springs, February 1987, Geodynamics Branch, Division of Earth Science and Applications, NASA, 45 pp.

National Research Council (NRC), 1990, Geodesy in the year 2000. Committee on Geodesy, Board on Earth Sciences and Resources, National Academy Press, Washington, D.C., 176 pp.

National Research Council (NRC), 1993, Solid-earth sciences and society. Committee on Status and Research Objectives in the Solid-Earth Sciences: A Critical Assessment, National Academy Press, Washington, D.C., 346 pp.

Nerem, R.S., and 19 co-authors, 1994, Gravity model development for TOPEX/POSEIDON: Joint gravity models 1 and 2. Journal of Geophysical Research (Oceans), v. 99, p. 24421-24447.

O'Loughlin, E.M., 1986, Prediction of surface saturation zones in natural catchments by topographic analysis. Water Resources Research, v. 22, p. 794-804.

Peterson, D.L., P. Popenoe, J.R. Gaca, and D.E. Karig, 1968, Gravity map of the Trinidad Quadrangle, Colorado. U.S. Geological Survey, Geophysical Investigations Map GP-638.

Phillips, J.C., 1993, Aeromagnetic investigations of hazardous waste sites. EOS, Transactions of the American Geophysical Union, v. 74, p. 220.

Pike, R., and G. Clow, 1981, The revised classification of terrestrial volcanos and catalog of topographic dimensions with new results of edifice volume. USGS Open File Report 81-1038, 40 pp.

Qidong, D., S. Fengmin, Z. Shilong, M. Li, T. Wang, W. Zhang, B.C. Burchfiel, P. Molnar, and P. Zhang, 1984, Active faulting and tectonics of the Ningxia-Hui Autonomous region, China. Journal of Geophysical Research, v. 89, p. 4427-4445.

Rapp, R.H., and N.K. Pavlis, 1990, The development and analysis of geopotential coefficient models to spherical harmonic degree 360. Journal of Geophysical Research, v. 95, p. 21885-21911.

Rowson, S., G.R. Courtney, and R.M. Hueschen, 1994, Performance of Category IIIB landings using C/A code tracking differential GPS. Navigation, v. 41, p. 127-143.

Rummel, R., and P. Teunissen, 1988, Height datum definition, height datum correction, and the role of the geodetic boundary value problem. Bulletin Géodésique, v. 62, p. 477-498.

Schwarz, K.P., O. Colombo, G. Hein, and E.T. Knickmeyer, 1992, Requirements for airborne vector gravimetry. Proceedings International Association of Geodesy Symposium, From Mars to Greenland: Charting Gravity with Space and Airborne Instruments, General Assembly of the IUGG, Vienna. New York: Springer Verlag, p. 273-283.

Schwarz, K.P., M.A. Chapman, M.E. Cannon, and P. Gong, 1993, An integrated INS/GPS approach to the georeferencing of remotely sensed data. Photogrammetric Engineering and Remote Sensing, v. 59, p. 1667-1674.

Seeber, G., 1993, Satellite Geodesy. Berlin: de Gruyter, 531 pp.

Simkin, T., L. Siebert, L. McClelland, D. Bridge, C. Newhall, and J.H. Latter, 1981, Volcanoes of the world: A regional directory, gazetteer, and chronology of volcanism during the last 10,000 years. Stroudsburg, PA: Hutchinson Ross Publishing Co., 232 pp.

Telford, W.M., L.P. Geldart, R.E. Sheriff, and D.A. Keys, 1990, Applied Geophysics, 2nd ed. New York: Cambridge University Press, 770 pp.

Ten Brink, U.S., Z. Ben-Avraham, R.E. Bell, M. Hassouneh, D.F. Colemen, G. Andreasen, G. Tibor, and B. Coakley, 1993, Structure of the Dead Sea pull-apart basin from gravity analyses. Journal of Geophysical Research, v. 98, p. 21877-21894.

Thomas, R.H., 1991, Polar Research from Satellites: A review. National Aeronautics and Space Administration, Washington, D.C.: Joint Oceanographic Institutions, 91 pp.

Thomas, R.H., K.C. Jezek, D. Wilson, W. Krabill, and K. Kuivinen, 1992, Changes in the surface topography of the Jacobshavn glacier. EOS, Transactions of the American Geophysical Union, v. 73, p. 175.

Topographic Science Working Group, 1988, Topographic Science Working Group Report to the Land Processes Branch, Earth Science and Applications Division, NASA Headquarters. Lunar and Planetary Institute, Houston, 64 pp.

Torge, W., 1989, Gravimetry. Berlin: de Gruyter, 465 pp.

U.S. Congress, Office of Technology Assessment, 1993, The future of remote sensing from space: civilian satellite systems and applications. OTA-ISC-558, Washington, D.C.: U.S. Government Print Office, 224 pp.

U.S. Congress, Office of Technology Assessment, 1988, Seismic verification of nuclear testing treaties. OTA-ISC-361, Washington, D.C.: U.S. Government Print Office, 152 pp.

U.S. Department of Transportation/U.S. Department of Defense, 1992, Federal Radionavigation Plan.   DOT-VNTSC-RSPA-92-2/DOD-4650.5, available from National Technical Information Service, Springfield, Va., 229 pp.

Watts, A.B., G.D. Karner, P. Wessel, and J. Hastings, 1985, Technical Report no. 4, CU-1-85. Office of Naval Research, Washington, DC.

Wells, D.E., N. Beck, D. Delikaraoglou, A. Kleusberg, E.J. Krakiwsky, G. Lachapelle, R.B. Langley, M. Nakiboglou, K.P. Schwarz, J.M. Tranquilla, and P. Vanicek, 1987, Guide to GPS Positioning. Fredericton: University of New Brunswick Graphic Services, 250 pp.